Clustering Techniques for Image Segmentation

Fasahat Ullah Siddiqui • Abid Yahya

Clustering Techniques for Image Segmentation

Fasahat Ullah Siddiqui
CQUniversity
Melbourne, Victoria, Australia

Abid Yahya
BIUST
Botswana International University of
Science and Technology
Palapye, Botswana

ISBN 978-3-030-81229-4 ISBN 978-3-030-81230-0 (eBook)
https://doi.org/10.1007/978-3-030-81230-0

This Springer imprint is published by the registered company Springer Nature Switzerland AG
The registered company address is: Gewerbestrasse 11, 6330 Cham, Switzerland

I dedicate this book to my family.
Fasahat Ullah Siddiqui
I dedicate this book to my family for their love, support, and sacrifice along the path of my academic pursuits.
Abid Yahya

Preface

This book consists of four chapters and is organized as follows:

Chapter 1 begins by explaining the concept of image segmentation and then comparing its working with image classification. Image segmentation's role in image processing represents and describes an image's different information into a simpler and understandable format. Usually, image segmentation is the initial step of a higher-level operation in an image processing system. Labeling and tracking of objects are examples of the higher-level operation that are usually performed after image segmentation. Image segmentation step can segment a data with one (a.k.a., 8 bits grayscale data) or three dimensions (a.k.a., 24 bits color data).

Chapter 2 discusses the hard and fuzzy partitioning clustering techniques and illustrates the dead center, center trapping, and outlier problems by using examples. Partitioning clustering segments images without any supervision or human interaction. The partitioning clustering process starts with a guess of a possible solution and updates the solution until there is no change in all cluster centroids. There are two different kinds of partitioning clustering techniques: (a) hard partitioning clustering and (b) fuzzy partitioning clustering. The hard partitioning clustering techniques have a hard membership function that updates the cluster-centroids using a similarity index. Often, the cluster centroid is trapped in the non-active region and becomes a dead center.

Chapter 3 discusses the possible enhancements in k-means clustering techniques to overcome its dead centers and trapped centroids problems and illustrates the working of two new enhanced versions of the k-means clustering techniques. Partitioning clustering is an unsupervised approach and can segment images in a less complex fashion. Thus, partitioning clustering is used in many digital image processing fields, for example, airborne and medical image processing. The k-means and fuzzy c-means clustering techniques are examples of popular hard and soft membership-based clustering techniques. The partitioning clustering techniques may have dead centers, trapped centroids, and outliers' sensitivity problems. Therefore, the partitioning clustering techniques do not always converge to optimum global location. The partitioning clustering techniques have been modified to overcome the mentioned problems.

Chapter 4 discusses the working of evaluation methods and the practical knowledge of the existing evaluation methods. The hard and soft clustering techniques have membership functions. The chief objective of these functions is converging the final solution at the optimum global location. This chapter discusses the existing quantitative analysis methods to demonstrate the segmentation performance of clustering techniques. In the earliest quantitative analysis methods of clustering techniques, the MSE (mean square error), inter-cluster variation, and VXB function have been generally used that measure the local cluster similarity only. As compared to the former quantitative analysis methods, the three modern methods measure the local cluster similarity and the global homogeneity of segmented images without any human interaction or predefined threshold settings.

Melbourne, Victoria, Australia Fasahat Ullah Siddiqui
Palapye, Botswana Abid Yahya

Acknowledgments

This book arose from the research work conducted by authors at Universiti Sains Malaysia, Malaysia. The authors would like to express their special gratitude and thanks to Universiti Sains Malaysia (USM); Botswana International University of Science and Technology (BIUST); Karakoram International University (KIU), Gilgit, Pakistan; University of Peradeniya, Sri Lanka; Sarhad University of Science & Information Technology (SUIT), Pakistan; City University of Science & Information Technology (CUSIT), Pakistan; and Shaheed Benazir Bhutto Women University Peshawar for giving us such attention, time, and opportunity to publish this book.

The authors would also like to take this opportunity to express their gratitude to all those people who have provided invaluable help in the writing and publication of this book.

Contents

List of Figures

List of Tables

About the Authors

Fasahat Ullah Siddiqui is currently working at Central Queensland University, Australia. His research education comprises a PhD degree with a specialization in remote sensing from Monash University, Australia. Before this, he received a bachelor's degree in Biomedical Engineering from Sir Syed University of Science and Technology (SSUET), Pakistan. He completed his master's degree in electronics engineering with a specialization in robotic-vision from Universiti Sains Malaysia (USM), Malaysia. He secured many awards throughout his education career, for instance, scholarships and studentship from SSUET, USM, and Monash University. In his ten years research career, he has published ten articles in internationally reputed conferences and journals, and his publications have received 131 citations in total. Besides this, both his h-index and i10-index are 5 (based on Google Scholar statistics). He reviews article papers for reputable journals like *Remote Sensing MDIP* and *IEEE Transaction on Geoscience and Remote Sensing*.

Abid Yahya began his career on an engineering path, which is rare among other researcher executives. He earned his bachelor's degree from the University of Engineering and Technology, Peshawar, Pakistan, in electrical and electronic engineering, majoring in telecommunication, and MSc and PhD degrees in wireless and mobile systems from Universiti Sains Malaysia, Malaysia. Currently, he is working at the Botswana International University of Science and Technology. He has applied this combination of practical and academic experience to a variety of consultancies for major corporations. Prof. Abid Yahya is a Senior Member of the

Institute of Electrical and Electronics Engineers (IEEE), USA, and a Professional Engineer registered with the Botswana Engineers Registration Board (ERB). He has published many research articles in numerous reputable journals, conference articles, and book chapters. He has received several awards and grants from various funding agencies and supervised several PhD and master's candidates. His recent four books are (1) *Emerging Technologies in Agriculture, Livestock, and Climate* by Springer International Publishing; (2) *Mobile WiMAX Systems: Performance Analysis of Fractional Frequency Reuse* published by CRC Press/Taylor & Francis; (3) *Steganography Techniques for Digital Images* by Springer International Publishing; and (4) *LTE-A Cellular Networks: Multi-hop Relay for Coverage, Capacity, and Performance Enhancement* by Springer International Publishing and are being followed in national and international universities. Prof. Yahya was assigned to be an external and internal examiner for postgraduate students. Prof. Yahya was invited several times to be a speaker or visiting lecturer at different multinational companies. He sits on various panels with the government and other industry-related boards of study.

Chapter 1
Introduction to Image Segmentation and Clustering

1.1 Digital Image Processing System

The digital image processing system has three phases, that is, image processing, image analysis, and decision-making. Figure 1.1 illustrates the main steps of the digital image processing system. Several tasks can be performed in the image processing phase, based on the computer vision application's requirements. In general, image enhancement, noise filtering, and image compression are implemented in the image processing phase to increase the image's quality and ease the following phases in digital image processing system. While in the image analysis phase, image segmentation and image representation play a fundamental role before applying images to higher level operations of the decision-making phase such as image classification and image matching. The prime focus in image segmentation is the clustering technique, and interested readers can refer to other literature to learn in detail other steps of the digital image processing system (Gonzalez et al., 2003; Shih, 2010).

1.2 Image Classification and Image Segmentation

Many of us get confused between image classification and image segmentation, and some consider both are similar because of implementing the segmentation techniques, like clustering for pattern recognition. In pattern recognition, clustering and classification are two significant techniques. The classification technique needs prior knowledge of class labels, whereas the clustering technique does not require such information. However, image segmentation and image classification are two different steps of the digital image processing system. Image classification's primary role is to recognize predefined objects in an image, while the role of image segmentation is very much limited to simplify an image into homogenous regions. Classification

F.U. Siddiqui, A. Yahya, *Clustering Techniques for Image Segmentation*,
https://doi.org/10.1007/978-3-030-81230-0_1

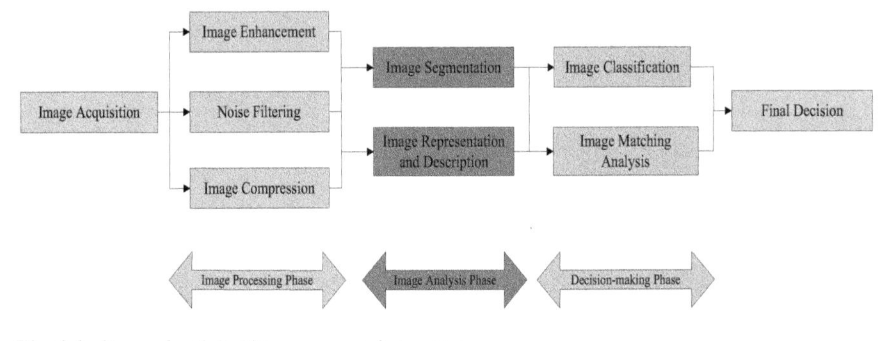

Fig. 1.1 Steps of a digital image processing system

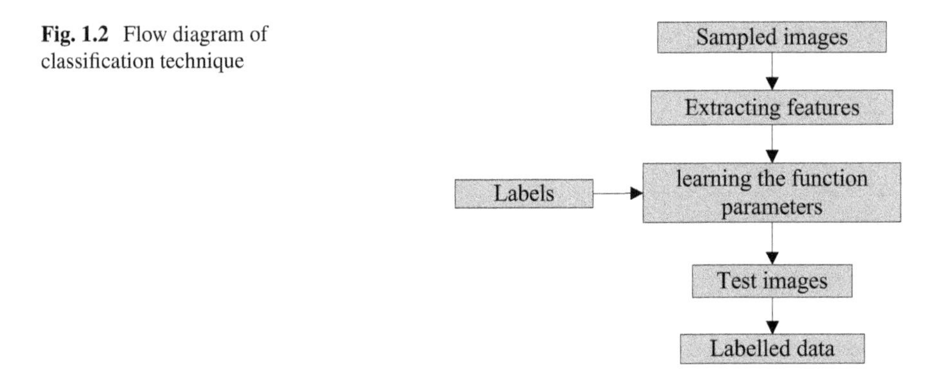

Fig. 1.2 Flow diagram of classification technique

is also a supervised technique that uses a series of input-output pairs (training images and classifying labels) to train a classification model with parameters. Later, the trained model is applied to a test image to estimate each pixel in the test image. Irrespective of the particular classification model, the essential steps of classification are as follows:

- Decide labels into which the test image is to be classified
- Estimate function parameters of a model using an algorithm on sample images
- Classify the test image into desire labeled classes using the trained model function

To better illustrate, a flow diagram of the classification technique is depicted in Fig. 1.2.

Figure 1.3 presents the working steps of the clustering technique. Unlike the classification technique, clustering does not need any prior information of object classes. It only divides the image pixels into regions based on specific attributes such as texture, gray scale, and color intensity (Khotanzad & Bouarfa, 1990).

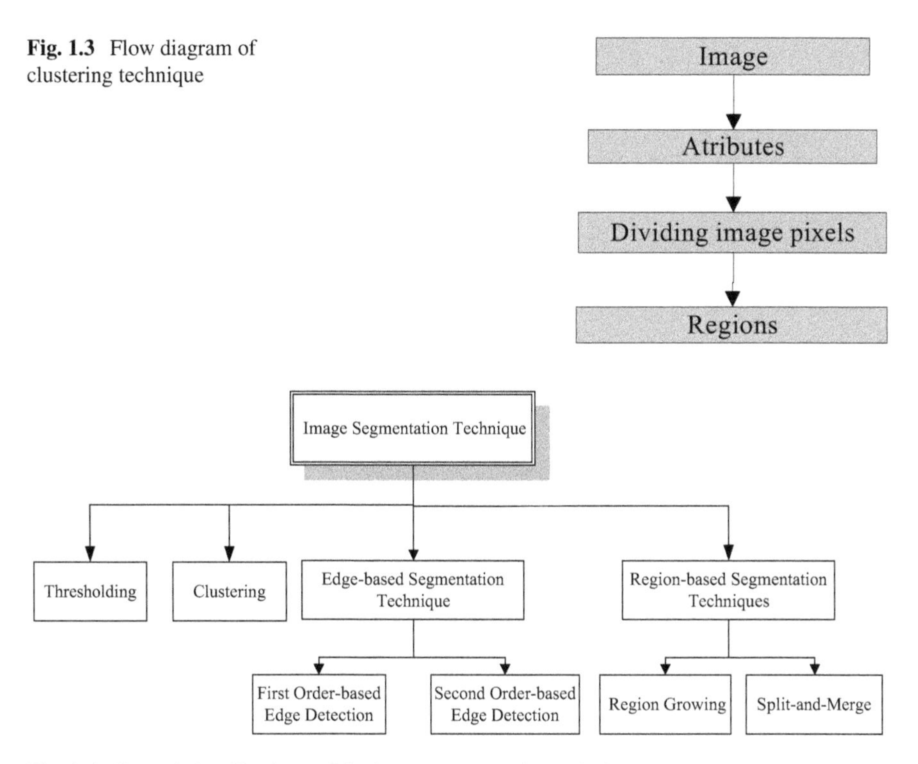

Fig. 1.3 Flow diagram of clustering technique

Fig. 1.4 General classifications of the image segmentation technique

1.3 Classification of Image Segmentation Technique

Several image segmentation techniques have appeared in the recent literature. Image segmentation techniques can be classified into four general categories: thresholding, clustering, edge-based segmentation technique, and region-based segmentation technique, as shown in Fig. 1.4. The remarkable modifications in the basic concept of thresholding, clustering, first-order-based edge detection, second-order-based edge detection region are growing, and split-and-merge techniques are discussed in the following sections.

1.4 Thresholding

Thresholding is a common segmentation technique. As shown in Fig. 1.5, it separates the object and background regions of an image through the selected threshold value. In other words, the pixels having an intensity value less than the threshold value and the pixels having intensity value more significant than the threshold value

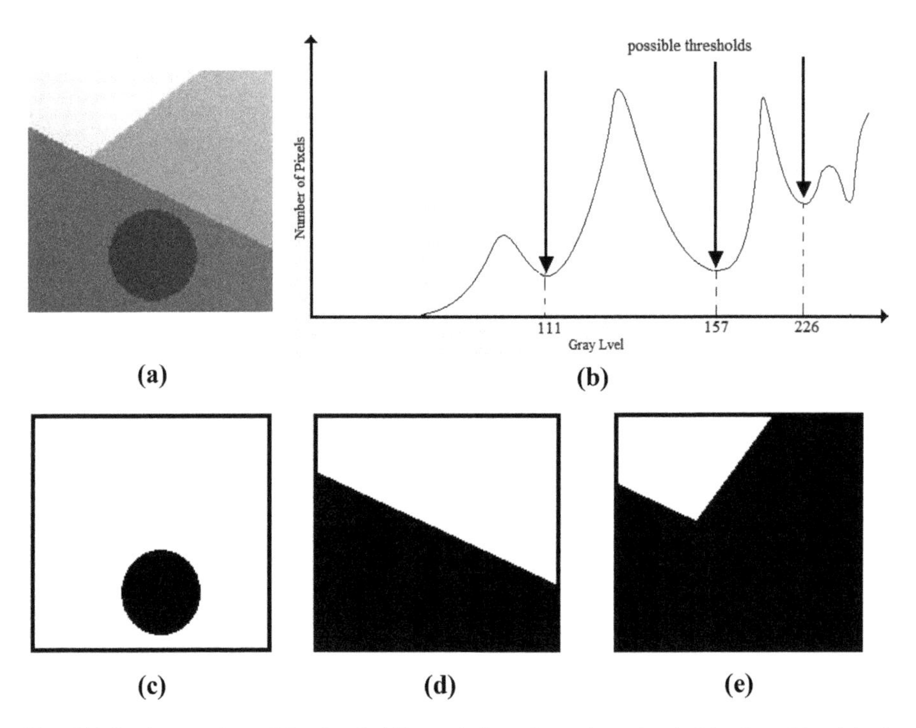

Fig. 1.5 Implementation of the thresholding technique for various threshold values (**a**) original image, (**b**) histogram of an image, (**c**) resultant image if the threshold is set to 111, (**d**) resultant image if the threshold is set to 157, and (**e**) resultant image if the threshold is set to 226

are grouped into two different regions. According to Sahoo et al., thresholding is mainly classified into local and global thresholding (Sahoo et al., 1988).

1.4.1 Global Thresholding

Global thresholding is the simplest method of thresholding technique. It compares the image pixels value with the threshold value. The pixels with more excellent value than the threshold are marked as 1 (first binary level), while the pixels having a value less than the threshold are marked as 0 (second binary level). It is expressed mathematically in Eq. 1.1, where $I(x, y)$ is image pixels and $B(x, y)$ is binary output after applying a T threshold.

$$B(x,y)=\begin{cases}1 & \text{if} I(x,y)>T\\ 0 & \text{if} I(x,y)\leq T\end{cases} \tag{1.1}$$

Therefore, thresholding methods are also known as binarization methods. Global thresholding is computationally simple and fast and works well if the image histogram is a bimodal shape. However, it fails if non-bimodal histogram images, such as noisy, intensity inhomogeneity, and low contrast images are segmented (Pham et al., 2000). These kinds of images are commonly produced by magnetic resonance imaging (MRI) and computerized tomography (CT) scans. Besides, different segmentation results obtained may have varying initial global threshold values. Therefore, the proper selection of the initial threshold value defines the accuracy of global thresholding. In this domain, many methods have been proposed, which are discussed in the following subsections.

1.4.2 P-tile Method

P-tile is the earliest existing method of global thresholding (Doyle, 1962). This method assumes that the object area's percentage ($P_b\%$) in a gray scale image is known before and brighter than the background. In this case, the threshold must be set as the gray level, which maps $P_b\%$ the image pixels into an object and pixels the image pixels into the background. This method will not work if the area of the object is unknown (Sahoo et al., 1988). Recently, this method has been combined with an edge detector to assist in the thresholding process for the image with an unknown object area (Samopa & Asano, 2009; Taghizadeh & Hajipoor, 2011).

1.4.3 Histogram Shape-Based Methods

1.4.3.1 Peak-and-Valley Methods

The mode method finds peaks and valleys in the histogram by measuring the local minima between two peaks or modes (Sahoo et al., 1988). This method cannot be applied to the image with unequal peaks (noisy image) and those with flat valley (low contrast image). Sezgin proposed a new method that finds the peaks and valleys by convolving the histogram function with a smoothing kernel (Sezgin, 2004). Therefore, the gray levels at which peaks start, end, and attain their maxima are estimated. By this process, the histogram is reduced to a two-lobe function. The threshold must be set somewhere between the terminating of the first lobe and the second lobe. Variations of this method have been proposed, where the cumulative distribution of the image histogram is first expanded in terms of Chebyshev polynomial functions and followed by curvature analysis (Boukharouba et al., 1985; Sezgin, 2004). The polynomial function with different degrees uses as polynomial curvature fit theory, where the objective is to find the coefficients that best fit the curve to the data. The critical points of the resultant curve, that is, minima or zero points, select as threshold points. The Gaussian filter is also used for smoothing the histogram. Tsai et al. employed it with curvature analysis to find peaks and valley. The instantaneous rate of change of angle is measured to locate the threshold point. However, this method only works well with the appropriate selection of the Gaussian filter's window size (Tsai, 1995). Large windows might over-smooth the histogram and skip the desire peaks.

On the other hand, more than desired number of peaks are obtained in the histogram, if too small a window is selected. Olivo considered the multi-scale analysis of the probability mass function by choosing the wavelet or smoothed histogram, which is the second derivative of the smoothing function, where the threshold is found. This threshold is adjusted by using the coarse-to-fine approach (Olivo, 1994; Sezgin, 2004). This adjustment is started from a threshold at the lowest resolution, linking all the thresholds in correspondence at high resolutions and backward update of their location (Olivo, 1994). The main disadvantage of these peaks-and-valley finding methods is their disregard for spatial information (Hemachander et al., 2007).

1.4.3.2 Convex Hull Methods

This method automatically finds the threshold by analyzing the histogram's concavity by constructing the convex hull; this is the smallest polygon envelope containing the histogram. The concavity can be defined by connecting the polygon envelope to the histogram heights. The threshold must be set at the histogram's deepest concavity (Rosenfeld & De La Torre, 1983). A variation of this method has been proposed, where the convex hull is constructed through the exponential of the histogram. The exponential characteristic of this method can indicate the upper concavities more precisely (Whatmough, 1991). Figure 1.6 illustrates the convex hull construction and exponential convex hull.

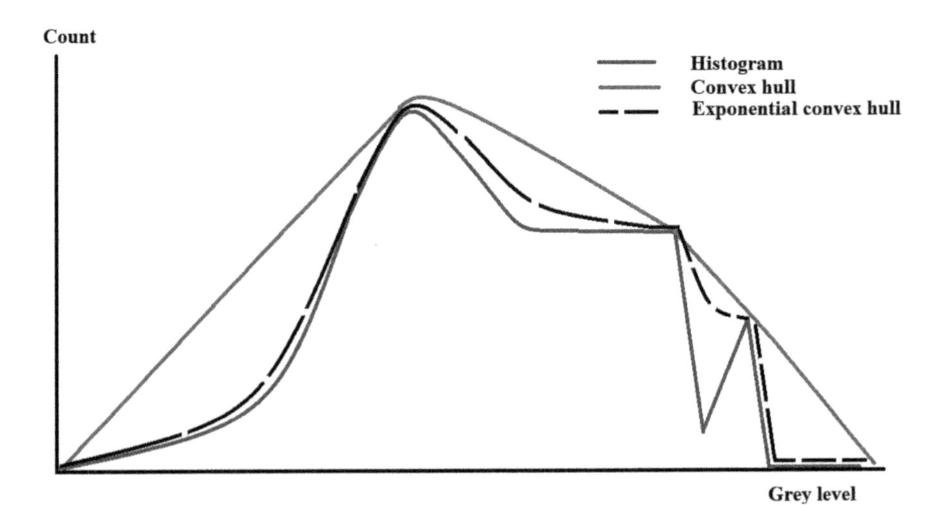

Fig. 1.6 The convex hull and the exponential convex hull

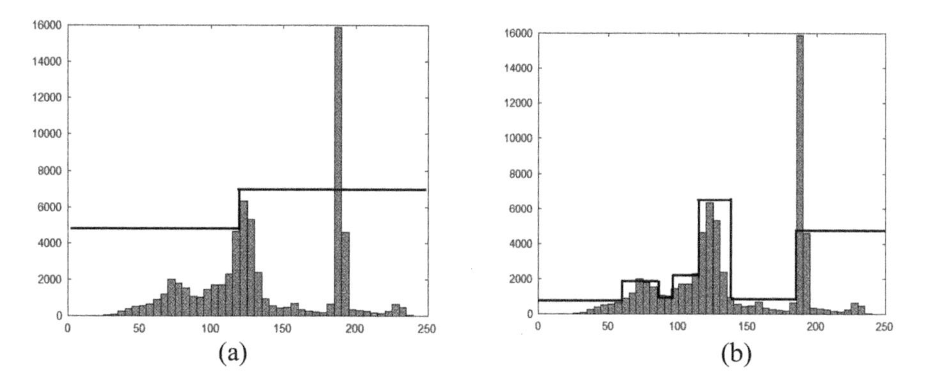

Fig. 1.7 Approximations of the histogram (**a**) at the first iteration and (**b**) after some iterations

1.4.3.3 Shape Modeling Methods

In shape modeling methods, the approaches based on the functional approximation of histogram have been studied widely. Ramesh et al. proposed a technique that uses the bilevel function representing the set of discrete values of bins (i.e., the number of pixels with a certain gray level) (Ramesh et al., 1995; Sezgin, 2004). The user sets the threshold value in this method. However, the histogram is mapped into two levels in the manual set threshold, as shown in Fig. 1.7a. Either side of the transition point or threshold point is a further map using the newly defined threshold point. The approximation process will be terminating at the minimum error value. The error may be measured by the variance or the sum of square error between the approximation function and histogram. Figure 1.7b represents the approximation function after the number of iterations.

Kampke & Kober, (1998) have generalized this shape approximation method. In their method, the approximation function (also known as quantization function) tends to assign breakpoints as a threshold where considerable variation in the histogram is observed (Sezgin, 2004).

Another approximation method is called the Prony analysis to estimate the histogram. It views the histogram as a mixture of all-pole distribution. Prony analysis method can linearly be extracting the sinusoid or exponential histogram by setting the linear equation to estimate the coefficient (predictor coefficient) that satisfies the histogram (Jinhai & Zhi-Qiang, 1998). Like the all-pole model method, the autoregressive method is also employed to build or extract the sinusoid (exponential histogram) using the variable lagged once or more times to estimate the new variable until it satisfies the histogram. In these methods, the variables or coefficients of the linear equation are calculated by using autocorrelation. The threshold is established as the minimum resting between its two-pole locations in the resulting smoothed histogram (Sezgin, 2004).

1.4.4 Clustering-Based Thresholding

By using this approach, the histogram data undergoes the clustering analysis. The given dataset is divided into two clusters with the initial parameter set based on assumption or guess. Here, the initial parameter may be the threshold position and the cluster class value. Based on the assumption (i.e., the optimum location is near the initial defined value), the clustering analysis searches for the optimum location. It will be stopped if the threshold position remains unchanged during the analysis or if it has attained minimum class variance. Other objective functions are used in different clustering analysis methods to search the optimal threshold position in histogram data. However, the clustering has some severe problems, that is, sensitive to initialization, class centroid (class representing point) will not be updated during the process, sensitive to outlier or noise, etc. Moreover, these problems are interrelated to each other and discussed in detail in the subsequent chapter.

1.4.4.1 Iterative Thresholding Methods

In early iterative thresholding methods, the initial threshold is set by assuming that the four corner pixels belong to the background, and the remainder are object pixels. The class means values are measured for the foreground (object) and background regions, respectively. The threshold value is updated iteratively by calculating the average of both class means (Ridler & Calvard, 1978). Here, the objective function is similar to the k-means objective (i.e., minimizes the class variance). This threshold updating process will stop when the error between the current and previous thresholds is very small. By completing the process, this method ensures that the Gaussian mixture distribution of histogram is grouped into two distinct classes with possible lowest class variance, and the threshold is set in between them. Trussell generalized this method by proposing a formula to measure the new threshold value (Trussell, 1979).

$$T_k = \frac{\sum_{g=0}^{T_k-1} g \times n(g)}{2\sum_{g=0}^{T_k-1} n(g)} + \frac{\sum_{g=T_k+1}^{Ng} g \times n(g)}{2\sum_{g=T_k+1}^{Ng} n(g)} \tag{1.2}$$

where, g is gray level, $n(g)$ is the number of pixels belonging to the gray level, Ng is the total number of gray levels and T_k is the threshold point.

The summation of these two quantities defined the new threshold value (Trussell, 1979). In other studies, the threshold is set by calculating the midpoint of the two peaks. Initially, the midpoint is measured by the pixel's average with maximum value and pixel with the lowest value. This midpoint will be updated in later iterations using the two-class peaks' mean (Sezgin, 2004; Yanni & Horne, 1994).

1.4.4.2 Minimum Error Thresholding Methods

Lloyd (1985) proposed a method that considers the equal variance Gaussian density functions for the two regions and minimizes the total misclassification error iteratively (Lloyd, 1985). The process initiates with the guess of optimal points, and the final solution will be obtained iteratively by minimizing the error of Gaussian functions. The threshold value is calculated by taking the average of two class means. Against the equal variance, the new Gaussian density function is introduced that minimizes the total misclassification error by fitting the Gaussian model to the histogram such that the histogram is clustered into two lobes (two normal distribution of histogram) with smaller overlapping (Kittler & Illingworth, 1986). However, the distributions' tails are truncated and they bias the actual model's mean and variance (Cho et al., 1989). This bias becomes noticeable when the two histograms are not distinguishable.

1.4.4.3 Ostu Clustering Thresholding Methods

Ostu method begins with the guess of final solution points. It searches iteratively by maximizing the weighted variance between the background and foreground classes to search for the optimal threshold point (Otsu, 1975). The probability theory measures the weights of the classes. It is believed that the variance between the classes (separability) is maximized by minimizing the within-class variance (similarity). The threshold is selected as an optimal histogram data threshold if the desired maximum variance between the classes is obtained. Unfortunately, the noise always occurs in practical applications and affects this method's accuracy (Lang et al., 2008). The one dimension of the histogram is insufficient to overcome this problem. Therefore, the two-dimensional (2D) Ostu method is proposed (Jianzhuang et al., 1991). The 2D histogram presents the original image pixels distribution, on the one side, and shows the average neighborhood image, on its other side. Therefore, the resultant threshold becomes the vector quantity, which improves the segmentation results. However, the computational cost will be increased. The diagonal areas of the 2D histogram represent the background and foreground. Most time is consumed for calculating the triangle areas. Lang et al. represent the 2D histogram in three integral images (i.e., pixel number integral image, original image intensity integral image, and average intensity integral image). Instead of using the vector value (i.e., two values) in all calculations, the three integral image values are directly substituted to calculate only the mean value (Lang et al., 2008).

1.4.4.4 Fuzzy Clustering Thresholding Methods

Fuzzy clustering with soft membership can reduce the initialization problem. With the assumed optimal points, the data are distributed partially to the classes, and new optimal points are calculated by the weighted (membership) sum of the data. This will continue until the difference between the membership function of current and previous iterations is minimized. The optimal points are considered as the desired points. On the other hand, Jawahar et al. formulate the distance function according to the kittler et al. minimizing function. It assumes that the optimal threshold point can be simplified by considering the normal distribution of object and background in a histogram (Jawahar et al., 1997). The memberships for background and object are calculated with randomly setting the initial threshold value by computing the mean value of both regions and updating membership according to a new function derived from the kittler et al. minimized function based on Gaussian function. It is the iterative process that will continue until there is no appreciable change for regions' membership. In other words, the weight based on the neighborhood is calculated (Yong et al., 2004). It is believed that the pixels intensity level with neighborhood information is prone to the noise effect. After initializing the membership and partially assigning the gray levels to the clusters, the probability of gray level along the neighboring clusters is calculated as an additional weight. The new degree is measured by multiplying the weight by the membership value. The threshold point is finally set at the clusters' midpoint after calculating the mean value of clusters and continuing the process until a very small change in the new degree is obtained. The threshold point is finally set at the midpoint of the clusters.

Pal & Rosenfeld (1988) highlighted the improper selection of the image's fuzzy region's threshold position. He argued that the threshold is the point or boundary, where the data is segmented in a crisp way (1 or 0). He regraded the fuzziness levels into 0 and 1 using subset theory (Pal & Rosenfeld, 1988). The cross-over point with a value of 0.5 is defined for a specific region in the histogram. The crisp version of an image is obtained by setting the membership to 1 if it is greater than the cross-over point in the gray level in the area of region or window and 0 for the rest. The fuzziness index calculates the average difference between the gray level and the obtained binary version. By varying the cross-point on the gray level in the window, the different fuzziness indexes are calculated. The cross-over point is selected as an optimal threshold where the fuzziness index is minimized. However, this geometrical method has no theoretical proof for choosing the constant for selecting the bandwidth of region or window. Murthy et al. were the first to theoretically choose the constant's value to define the region (Murthy & Pal, 1990). They set the known maximum value of subset "0" as the minimum limit and the known minimum value of subset "1" as the region's maximum limit. Therefore, their average is equal to the gray level, that is, cross-over point. In another study, Huang & Wang (1995) measure the fuzziness index using the mean or median operator. The fuzziness index is calculated with the assumption that set "0" has no elements and others have all, and the threshold is at 0 of gray level on the histogram. Iteratively, sift the threshold point and calculate the fuzziness index (i.e., within the range of 0.5 to 1). The histogram's point is to select as the optimum threshold point, where the fuzziness index should be minimized. Besides, the fuzzy range is defined for measuring fuzziness index that equals to or less than tolerance; this ensures that the threshold lies on the deepest valley.

$$tolerance = \min fuzziness + (\max fuzziness - \min fuzziness) \times \alpha\% \quad (1.3)$$

where α is the specified value $0 \leq \alpha \leq 100$

Another approach, that is, fuzzy type II (fuzzy subset II or ultra-fuzziness) theory, is proposed to search for the optimal threshold point (Tizhoosh, 2005). After initializing, the membership function calculates the fuzziness subset II by dividing the gray level membership using its intensity value. The fuzziness subset II is itself fuzzy, which can model (footprint) the uncertainties in the histogram. The upper and lower limit of fuzzy subset II is calculated for an individual gray level and their average difference becomes ultra-fuzziness. The histogram location with maximum ultra-fuzziness is set as a threshold point. This theory was generalized and reformed by the histogram in 2D space. One is the gray level, and the other is local information (i.e., the number of neighboring pixels with the same gray value to the gray level) (Xiao et al., 2011). In addition, the histogram is stretched on a gray value scale by obeying the Weber law (i.e., the difference between the two levels or regions will be maximized). After this ultra-fuzziness at different gray levels is calculated by fuzzy set type II, the ultra-fuzziness with maximum value becomes the threshold value.

1.4.5 Entropic Thresholding Methods

Entropy converts the histogram data into the binary stream by exploiting the redundancies in the gray level's statistical distribution (l) to reduce as much as possible the size of the binary stream. The maximization of the entropy of the threshold image is used in the entropic thresholding-based methods. Assuming the prior entropy of gray-level histogram, Pun et al. determined the optimal threshold by maximizing the posterior entropies associated with a gray scale image's brighter and dark pixels (Pun, 1981; Sahoo et al., 1988).

$$H_b = -\sum_{i=0}^{t} p_i \log_2 p_i \quad (1.4)$$

$$H_d = -\sum_{i=t+1}^{l-1} p_i \log_2 p_i \quad (1.5)$$

$$H = H_b + H_d \quad (1.6)$$

In his next work, Pun et al. defined a coefficient (α) to estimate the threshold.

$$\alpha = \frac{\sum_{i=0}^{m} p_i \log p_i}{\sum_{i=0}^{l-1} p_i \log p_i} \quad (1.7)$$

Here, m is the smallest inter such that

$$\sum_{i=0}^{m} p_i \geq 0.5 \tag{1.8}$$

Entropic thresholding also used the 2D histogram instead of the one-dimensional histogram, where the spatial and spectral intensity is accommodated. Interested readers can read more by Sahoo et al. (1988).

1.4.6 Local Thresholding

Local thresholding is an adaptive thresholding technique that is comparatively less sensitive to noise. It compares each pixel individually in its local neighborhood pixels. The neighborhood pixels are assigned by convolving a $w \times w$ window on an image. The image pixel $I(x,y)$, that is, locates at the window center, is marked as 0 if it is less than or equal to the locally measured threshold value. Thus, an individual local threshold $T(x,y)$ is measured to generate the $B(x,y)$ binary result for each image pixel. It can be expressed mathematically by,

$$B(x,y) = \begin{cases} 0 & \text{if} I(x,y) \leq T(x,y) \\ 1 & \textit{otherwise} \end{cases} \tag{1.9}$$

The local threshold can be measured in different ways. Bernsen measured a local threshold by using minimum and maximum intensity in a local window. Therefore, the local threshold is set at the midrange of intensities in a local window (Bernsen, 1986).

$$T(x,y) = 0.5\left(I_{\max(i,j)} + I_{\min(i,j)}\right) \tag{1.10}$$

A contrast constant is used to simplify the measurement further and speed up local adaptive thresholding. If the contrast is less than 15, then the pixel belongs to its neighboring pixels' class (foreground or background). This method performs best at window size 31.

$$C(i,j) = \left(I_{\max(i,j)} + I_{\min(i,j)}\right) \geq 15 \tag{1.11}$$

A mathematical calculation, such as mean and variance of pixels intensity in a window, was also used to compute a local threshold.

$$T(x,y) = m(x,y) + ks(x,y) \tag{1.12}$$

or

$$T(x,y) = m(x,y)\left[1 + k\left(\frac{s(x,y)}{R} - 1\right)\right] \tag{1.13}$$

In Eqs. 1.12 and 1.13, R is the maximum value of the standard deviation (i.e., 128 for gray scale image) and k is positive value constant (i.e., in the range of [0.2,0.5]), while $m(x, y)$ is mean, and $s(x, y)$ is the standard deviation of the image pixels present in a $w \times w$ window centered around the pixel $I(x, y)$. In other words, the local threshold value is set based on the pixel intensities contrast in a window. Local threshold measurement takes considerable time for binarization of an image. One way to speed up the binarization process of local adaptive thresholding is by using a small size window. The small window size reduces the computational cost of the threshold calculation. Another way is to measure the local mean and standard deviation in a window of any size using the integral sum image technique before a local mean and standard deviation calculation. Viola and Jones implemented an essential image for computer vision (Viola & Jones, 2004). Consider I_g is an integral image of an image I where the pixel intensity I_g is measured by adding the intensity of all pixels above and to the left of that pixel position in an image I. It can be expressed mathematically by,

$$I_g = \sum_{i=0}^{x}\sum_{j=0}^{y} I(i,j) \tag{1.14}$$

The local mean and standard deviation for any window size can be computed using two addition and one subtraction operations instead of summing pixel values in a local window (please refer to Eqs. 1.15 and 1.16 for mathematical representation). Therefore, integrating the integral image approach in the local adaptive threshold makes it flexible to use any window size without increasing computation cost.

$$m(x,y) = \frac{\left(I_g\left(x + w/2, y + w/2\right) + I_g\left(x - w/2, y - w/2\right)\right) -}{\left(I_g\left(x + w/2, y - w/2\right) + I_g\left(x - w/2, y + w/2\right)\right)} \tag{1.15}$$

$$s^2(x,y) = \frac{1}{w^2}\sum_{i-x-w/2}^{x+w/2}\sum_{j-y-w/2}^{y+w/2} I(i,j) - m^2(x,y) \tag{1.16}$$

The computational cost of the locally adaptive thresholding can be further decreased by replacing the standard deviation in Eq. 1.12 with the mean absolute deviation. As shown in Eq. 1.18, the mean absolute deviation is directly computed from the local mean.

$$\partial(x,y) = I(x,y) - m(x,y) \tag{1.17}$$

$$T(x,y) = m(x,y)\left[1 + k\left(\frac{\partial(x,y)}{1-\partial(x,y)} - 1\right)\right] \tag{1.18}$$

Several other methods use global image information while measuring a local threshold, and interested readers can view the text published in (Ismail et al., 2018) for more details.

1.5 Clustering

Clustering is the pixel-based process that organizes the raw data into clusters or groups whose members (pixels) are similar in some sense (Fig. 1.8). This can be achieved by using both the unsupervised and supervised approaches. The unsupervised clustering approach classifies the image into a subgroup without any learning of data. It uses the predefined objective function to partition the image into disjoint classes or clusters. The pixels within a cluster are similar as possible, and the pixels among the cluster are dissimilar as possible. The unsupervised clustering approach's main advantage is inherent in its simplicity and ease of implementation (Hamerly & Elkan, 2002; Vantaram & Saber, 2012). However, the clustering technique's accuracy highly depends on the initialization (i.e., initial clusters centroid positions). Furthermore, the clustering process becomes more difficult with the increase of dimensionality of space features, e.g., color image and texture.

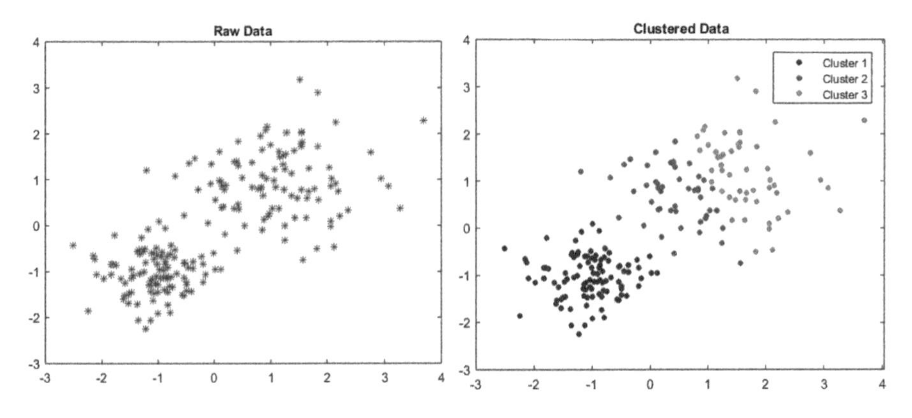

Fig. 1.8 Raw data and clustered data

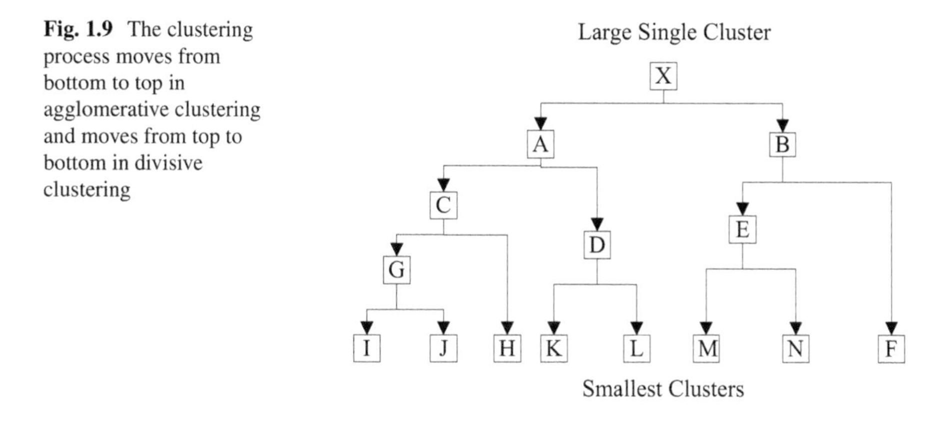

Fig. 1.9 The clustering process moves from bottom to top in agglomerative clustering and moves from top to bottom in divisive clustering

1.5.1 Hierarchical Clustering

Hierarchical clustering is a connectivity-based clustering method. It organizes the data into a tree structure or dendrogram, according to the proximity matrix. The dendrogram's root node represents the entire dataset, and each leaf of a dendrogram is depicted as a data object. Hierarchical clustering has two versions, namely, agglomerative and divisive clustering. In the former clustering algorithm, the entire data is considered a cluster, and the clustering process merges them in a single cluster.

On the other hand, the divisive clustering proceeds in the opposite direction. Initially, entire data belongs to a cluster, and the process splits it until all clusters are singleton. Figure 1.9 illustrates the working of agglomerative and deceives clustering methods. Due to the very high compositional time, divisive clustering is rare for segmentation (Xu & Wunsch, 2005).

Based on the different existing definitions of distance, the agglomerative clustering method is mainly classified into single linkage, complete linkage, average linkage, median linkage, centroid linkage, and ward's linkage or minimum variance linkage (Table 1.1) (Olson, 1995). Commonly, the agglomerative clustering (AC) methods are not adaptive to noise and outliers (Almeida et al., 2007). Furthermore, they are susceptible to clusters' shape and size (Almeida et al., 2007; Zhao et al., 2005). The major drawback of these algorithms is that the previous misclassification can't be fixed in the future and take a long time to compute the extensive data (Xu & Wunsch, 2005; Zhao et al., 2005). For example, in Fig. 1.9, if some elements are misclassified and become a part of "*B*" subset instead "*A*", then these elements have no chance to become a part of A in a later process. Recently, many modifications have been made to overcome the abovementioned limitation of HC methods. Some of them are ROCK, CRUE, Chameleon, and BRICH.

ROCK measured the cluster similarity index based on the number of points from different clusters with a common neighbor. This similarity index is used to link the cluster with the other cluster and ensure that the best pair of clusters are merged in each step or iteration. This approach can effectively reduce the outlier sensitivity in

Table 1.1 Distance calculation for different approaches

Classification of Agglomerative clustering method	Distance between the two clusters is equal to:
Single linkage	The shortest distance from any one member of one cluster to any member of the other cluster.
Complete linkage	The most significant distance from any one member of one cluster to any member of the other cluster.
Average linkage	The average distance from any one member of one cluster to any member of the other cluster.
Median linkage	The distance from the median of one cluster to median of the other cluster.
Centroid linkage	The distance from the centroid of one cluster to centroid of the other cluster.
Ward linkage	The distance from the centroid of one cluster to centroid of the other cluster. Here, the centroid is the point with the minimum variance within the cluster.

extensive data (Almeida et al., 2007; Guha et al., 2000). Like ROCK, the random sample strategy is used by CURE to handle the large data. It combines the complete and single linkage methods for choosing more than one representative of a cluster. In each step or iteration, the two clusters with the closest pair of representative data points or objects are merged. Thus, the scattered points are shrunk toward the cluster mean point and become less sensitive to outliers (Guha et al., 2001).

On the other hand, Chameleon clustering uses the novel approach. It measures the similarity between each pair of clusters by looking both at their relative interconnectivity and their relative closeness. Chameleon's clustering selects to merge the pair of clusters for which both factors are high; that is, it selects to merge clusters that are well inter-connected and close together with the internal interconnectivity and closeness of the clusters. This allows Chameleon to handle the high feature space data by reducing the outlier effect (Karypis et al., 1999).

Noticing previous hierarchical clustering algorithms' restriction, BRICH has been proposed with a new data structure called cluster feature tree. The cluster feature tree stores summaries of each data point in clusters, much smaller than the original data. This allows it to quickly compute the multidimensional data without scanning whole data in its original format (Xu & Wunsch, 2005; Zhang et al., 1997). Initially, the cluster feature of data objects or leaf nodes is calculated, which contains necessary information (i.e., several data objects, the linear sum of data objects, and the square sum of the data objects) intra-cluster distance. Through the predefined parameters, that is, T as a threshold and B as a possible number of entries (sub-clusters), the capacity of root nodes for a particular level is determined to absorb the subclusters in them. The values of T and B are iteratively refined such that the outliers will not be a part of a node, and maximum numbers of similar leaf nodes are grouped to their closet root node. This process will be performed at each level until the level with a single root node is obtained. Therefore, the root nodes' size remains the same at an individual level, which is practically not right.

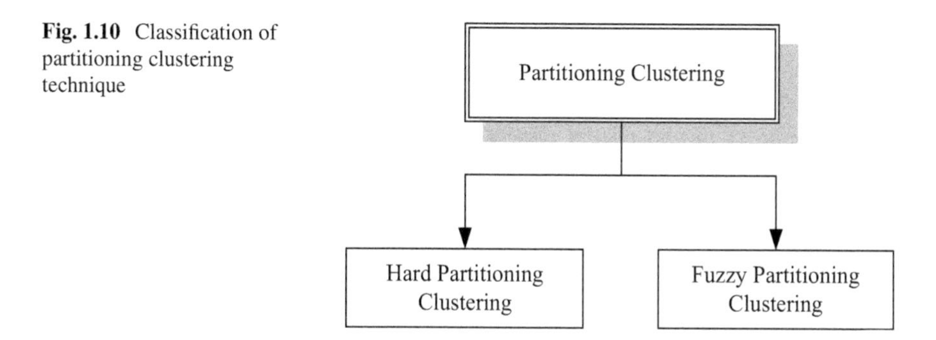

Fig. 1.10 Classification of partitioning clustering technique

1.5.2 Partitioning Clustering

Partitioning clustering starts with an initial partition of data into several clusters and then tries to swap data from one group to another iteratively. The objective function is optimized. Generally, the partitioning clustering is classified into two categories, that is, the hard partitioning clustering and fuzzy partitioning clustering, as shown in Fig. 1.10.

1.5.2.1 Hard Partitioning Clustering

K-means clustering is generalized of the Centroid Voronoi Tessellation (CVT) method. According to certain norm distances, the data is assigned to their closest sites or clusters, and the element of clusters is equivalent to their center mass value (Du & Gunzburger, 2002). K-means is an iterative method to optimize the function, that is, minimized the inter-cluster variance (within-cluster variance). It is initialized by randomly selecting the cluster's center value called the centroid. The hard membership for the pixels is calculated. Its value becomes one for a pixel to the closest cluster and zeroes in other clusters. Therefore, hard membership restricts the pixels to belong to only one cluster. The new centroid value is obtained by taking the mean of cluster pixels. This process is continued until no changes are observed in the value of the centroid.

The centroid trapped at local minima and dead centroid (cluster with no elements or pixels) has been addressed by Mashor, 2000. He proposed the Moving K-means (MKM) with a moving concept for eliminating the problems relates to K-means (Mashor, 2000). The fitness (i.e., variance within the cluster) value for a cluster is measured. It moves the cluster elements with the highest fitness value to the cluster with the lowest fitness value cluster. This concept of moving elements will ensure that the clusters' variance is similar, and the solution is converged to local minima. However, the intra-cluster variance (variance within the cluster) for clusters will not be attributed to the elements transferred to the wrong cluster. Adaptive Moving K-means (AMKM) proposed the solution and transfers the elements from the cluster with the highest fitness value to the nearest cluster (Isa et al., 2009). This approach reduces the problems, i.e., centroid trapping at the non-active region and dead centroid. The enhanced versions of AMKM and MKM called Enhanced Moving K-means EMKM are proposed (Siddiqui & Mat Isa, 2011). The key difference in these two versions is the range of elements that will be transferred from the cluster with the highest fitness value to the nearest cluster. This highlights the wrong selection of an element in AMKM and MKM. This not only overcame the above-listed problems but also resolved the stability of the system.

1.5.2.2 Fuzzy Partitioning Clustering

Unlike hard partitioning clustering, which assigns the pixel to a single cluster, fuzzy portioning clustering allows the pixels to belong to all clusters partially. Fuzzy C-means (FCM) clustering is one of the earliest methods that partially assigned the pixel to clusters (Bezdek, 1980). It employed the soft membership function that derived from its objective function.

$$FCM\left(X,C\right)=\sum_{i=1}^{N}\sum_{j=1}^{k}u_{ij}x_i-c_j^{\,2} \tag{1.19}$$

where k are several clusters, N is several pixels, and U_{ij} is membership function which is defined by:

$$u_{ij}=\frac{x_i-c_j^{\,2/(1-m)}}{\sum_{j=1}^{k}x_i-c_j^{\,2/(1-m)}} \tag{1.20}$$

The cluster centroid value is measured by averaging the pixels value with different degrees specified by the membership function. This process is continued until the difference between two iterations' membership function should be less than the predefined value in the range of 0 to 1. At the end of the process, it is assumed that the process will be converged to the optimum location. The fuzzy concept allows overlapping clusters, reducing initialization sensitivity and making FCM sensitive to outliers (Dixon et al., 2009; Yang et al., 2004). According to Kersten (1999), the Euclidean distance of membership formula is a highly sensitive outlier. It puts considerable weight on outlying pixels that pull the cluster's centroid from its optimum location (Kersten, 1999). To overcome this, FCM was generalized, and Lp Norm Fuzzy C-Means (Lp Norm FCM) was proposed. It uses the Lp Norm distance (i.e. $0 < p < 1$) instead of Euclidean distance (Norm 2) in the membership function (Hathaway et al., 2000). It works better than the FCM if the exact value of p is known for the particular data or image.

Wang et al. (2004) also address Euclidean distance sensitivity for the data with more than one feature space and proposed the feature-weighted Euclidean distance (Wang et al., 2004). Feature-weighted Euclidean distance replaces the Euclidean distance of FCM for reducing the sensitivity of Euclidean distance to the outlier. The feature weight is calculated using mean similarity indexes on Euclidean space and an image's feature space. It assumes that the weighted feature function is minimized, which attained the 0 or 1 value for similarity index on feature space. This minimized feature weight value is used to calculate the weighted distance in every repetition of the process. However, it is a complex method to measure the feature weight (Hung et al., 2008). He proposed bootstrap as a statistical approach to measuring the feature weight. In this statistical approach, the normalized value of variability is measured to define the feature weight.

In another approach, the distance is calculated between the centroid and weighted pixels. With this modification in conventional FCM, the Lagrange multiplier method is introduced to enhance the objective function of FCM (Chih-Cheng et al., 2011). The Lagrange multiplier is added or subtracted in function such that the objective function is minimized to zero, and the optimum location is obtained. Therefore, the two extra functions, that is, Lagrange multiplier and mean of weighted pixels, are calculated before measuring the membership function. The cluster fitness checking concept applies to FCM, and if the fitness value of clusters is highly different from each other, the membership will come from the cluster with a high fitness value to the nearest cluster. However, the trapped centroid at local minima is unsolved by it.

Later, the neighboring pixels (spatial) information calculates the pixels' membership value (Chuang et al., 2006). The window is defined for considering the specific number of neighboring pixels. The spatial function that measures neighboring pixels' influence has become one if the window is homogenous (pixel has similar neighboring pixels). This work is generalized, and the spatial function is also included in distance calculation (Huynh Van & Jong-Myon, 2009). This method assumed that the effect of salt and paper and Gaussian noise on segmentation is reduced by using the proper window size and the constant value that controls the influence of neighboring pixels on the membership of an individual pixel.

1.6 Region-Based Segmentation Techniques

1.6.1 Region-Growing Technique

Region growing is a technique of extracting the region in an image connected according to some predefined criteria (i.e., gray intensity, color intensity, or edges in the image). This basic approach begins from a predefined seed pixel in the predefined region seed that is to be segmented. The region-growing process includes all the neighboring pixels connected to predefined seed pixel based on predefined criteria. The process continues until all pixels have been considered or the produced region cannot be grown anymore. Figure 1.11 illustrates the region's implementation, while Fig. 1.12 shows an example of region-growing application for the image segmentation process.

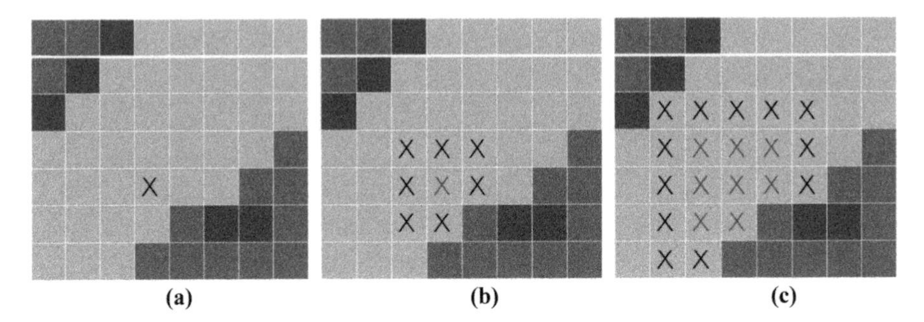

Fig. 1.11 Implementation of region-growing technique (**a**) location of initial seed pixel, (**b**), and (**c**) the growing process of seed pixel to form a region

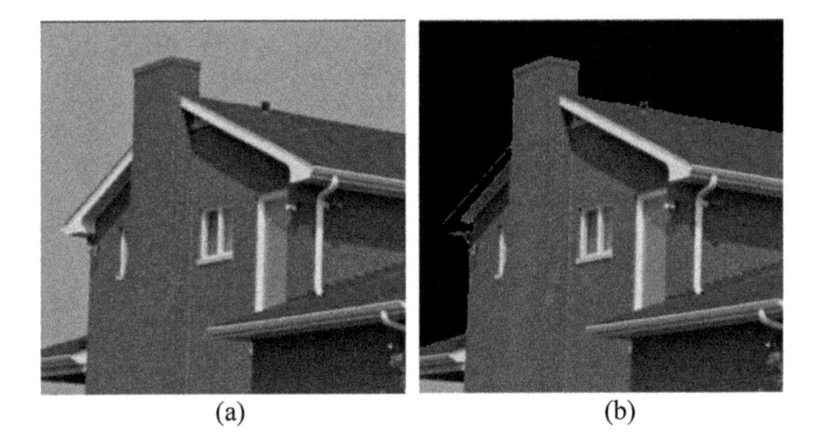

Fig. 1.12 The example of region-growing application (**a**) original image (**b**) segmented image (where, the extracted background region is converted into black color)

The region-growing technique offers several advantages: simple implementation, low computational time, and the capability to provide parallel clustering. However, it is sensitive to noise and variation of the intensities in an image, leading to segmented regions with small holes or disconnected (Sun & Bhanu, 2009). Besides, the region-growing technique requires manual interaction to obtain appropriate initial seed pixels, which is too subjective and time-consuming, especially for a complex image (i.e., medical images).

In terms of applications, the region-growing technique is often applied for medical images, where the interested region will be grown and distinguished from the surrounding region called background. Pohle and Toennies (2001) have successfully employed the region technique to segment out the kidney cortex, the liver, and the passable lumen in an aortic aneurysm from CT scan images. In 2005, Mat Isa applied the region growing on Pap smear images to segment the cell's cytoplasm and nucleus. In addition, Mazonakis et al. (2001) proposed the region-growing technique for prostate cancer segmentation on CT images. These applications could highlight certain regions on medical images for an easier screening process by a doctor. Other than that, the region-growing process has also been applied in remote sensing (Abad et al., 2000, Dare, 2005), satellite images (Bins et al., 1996, Derrien & Le Gléau, 2007, Anil & Natarajan, 2010), and face recognition (Jin et al., 2007).

1.6.2 Split-and-Merge Technique

The split-and-Merge technique is a recursive partitioning technique, where each cluster (as a root of a tree) is firstly split along its co-ordinate axes to yield four sub-clusters. If the obtained sub-clusters are non-homogeneous (i.e., not equal to the predefined threshold value), then these sub-clusters are further split into four sub-squares until all pixels have been considered. On the other hand, if at least two out of these four sub-squares are homogenous, they can be merged to form a bigger cluster. This process continues recursively until no further splitting or merging process is possible (Pavlidis & Horowitz, 1974; Horowitz & Pavlidis, 1976). The abovementioned process could be illustrated, as shown in Fig. 1.13.

Similar to the thresholding and region-growing techniques, the split-and-merge technique is commonly applied as a segmentation tool due to its simple implementation. Although its performance is comparable with others, it suffers several limitations. The split-and-merge process is laborious and requires high computational time as the process must be applied to the whole image (Faruquzzaman et al., 2009).

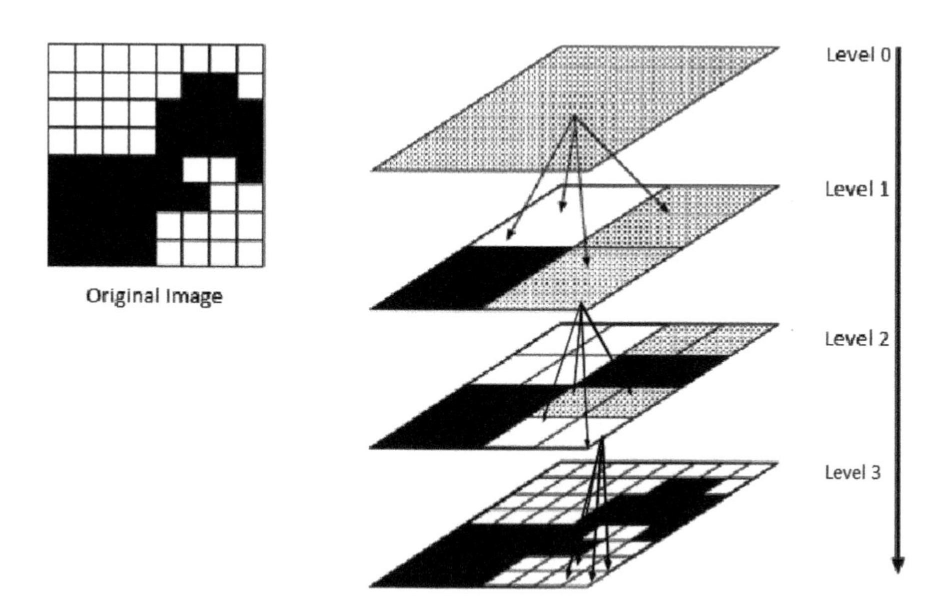

Fig. 1.13 Implementation of the split-and-merge technique. (Siddiqui, 2012)

Higher processing time is required if bigger image size is under consideration. Other limitations are its segmentation performance subjectively, which depends on the settling of the initial threshold value (Faruquzzaman et al., 2009), the final solution could probably converge to the insignificant optimum location (Faruquzzaman et al., 2008) and inefficient to extract certain shapes of regions (specifically for the irregular shape) with smooth boundary and edges (Faruquzzaman et al., 2008).

Based on the latter limitation, the split-and-merge technique is only applied for a specific application (Faruquzzaman et al., 2009). One of the applications is the segmentation of the structure of building in the remote sensing images (Rau & Chen, 2003, Khoshelham et al., 2005). In that application, the structural information is essential for building maintenance and urban planning. Besides, the split-and-merge technique is also applied in medical imaging (Manousakas et al., 1998; Nedzved et al., 2000), satellite imaging (Lucieer & Stein, 2002, Devereux et al., 2004), and face recognition (Sengupta et al., 2000, Meethongjan et al., 2010) areas.

1.7 Edge-Based Segmentation Techniques

The image can be segmented into the region by detecting boundaries based on discontinuity in intensity (gray levels). The first step of boundary detection is to detect points, lines, and edge discontinuities in an image. Points and lines are those discontinuities detected by simply convolving the window or masking an image. An example is shown in Fig. 1.14, where a 3 × 3 mask is depicted, and each element of the mask has a certain value. When it is convolving on an image, it computes a sum of the elements' product with an image's intensity.

$$M = \sum_{i=1}^{9} E_i I_i \tag{1.21}$$

For detecting points, the center element is set as a positive integer, and the rest of the elements are set as a negative integer. In other words, the center value is differentiating with the neighboring value of the mask on an image. A threshold is defined such that if $M > T$ then the image pixel location at the mask center element is detected as a point.

E_1	E_2	E_3
E_4	E_5	E_6
E_7	E_8	E_9

−1	−1	−1
−1	8	−1
−1	−1	−1

Fig. 1.14 Value of all elements of the 3 × 3 mask

-1	-1	-1
2	2	2
-1	-1	-1

-1	2	-1
-1	2	-1
-1	2	-1

-1	-1	2
-1	2	-1
2	-1	-1

2	-1	-1
-1	2	-1
-1	-1	2

Fig. 1.15 Value of all elements of the 3 × 3 mask at 0, 45, 90, and -45°

Fig. 1.16 Intensity graph of four main types of edges found at any boundary. (**a**) Step edge (abrupt change in intensity), (**b**) Ramp edge (gradual change in intensity), (**c**) Roof edge (gradual change in intensity), (**d**) Line edge (abrupt change in intensity)

Similarly, the lines are detecting in an image. Few changes in elements of a mask make it capable of detecting lines. For example, in Fig. 1.15, horizontal, vertical, inclined (+45°), and declined (−45°) elements of the mask are set as positive integers. The rest of the elements are set as negative integers to detect horizontal, vertical, and 45° lines in an image. The pixel may be detected as part of a multiple line-type. This is solved by comparing the *M* value of each line-type mask with others, and line-type has the highest *M* value, which is set as line-type of the pixel.

Mainly four types of edges may be observed in an intensity image, such as step, ramp, roof, and line edges (Fig. 1.16). Lines and point detection methods are not robust to detect all types of edges. Therefore, several advancements have been made in points and line detection methods to detect all kinds of edges.

1.7.1 First-Order Derivative Edge Detection

A first-order spatial derivative of intensity change produces a positive value. In the first-order derivative edge detection technique, the gradient is measured at two orthogonal directions in an image. Then a set of directional derivatives is computed to detect edges. In other words, the edge gradient $G(x, y)$ is measured along a line normal to the edge slope at image pixels $I(x, y)$.

$$G(x,y) = \frac{dI(x,y)}{dx}\cos\theta + \frac{dI(x,y)}{dy}\sin\theta \tag{1.22}$$

Then, the amplitude or magnitude of the edge gradient is expressed by

$$G(i,j) = \left|G_r(i,j)\right| + \left|G_c(i,j)\right| \tag{1.23}$$

Here, $G_r(i,j)$ is row gradient and $G_c(i,j)$ is column gradient. The direction of the edge gradient concerning row gradient is measured by:

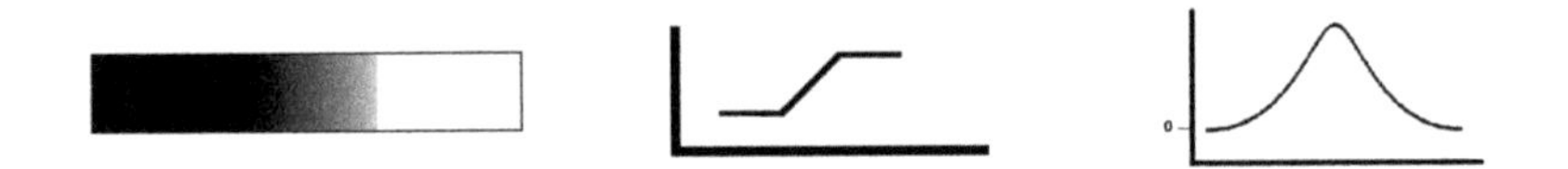

Fig. 1.17 Intensity graph of first-order derivative ramp edge gradient

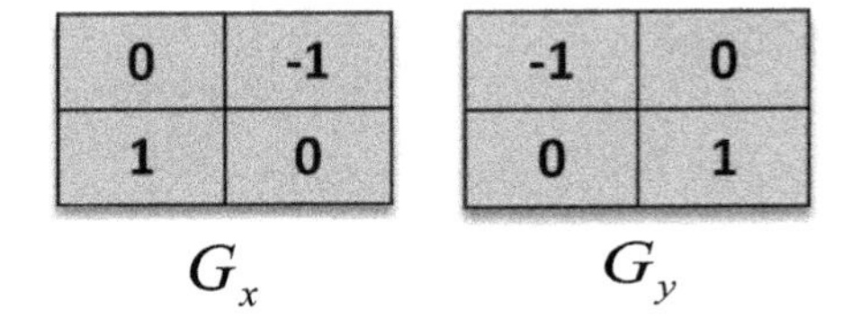

Fig. 1.18 Gradient value of all elements of the 2 × 2 mask/operator at x and y directions

$$\theta(i,j) = \arctan\left(\frac{G_c(i,j)}{G_r(i,j)}\right) \tag{1.24}$$

As shown in Fig. 1.17, the magnitude of the first-order derivative ramp edge gradient changes from 0 to 1 if a change in intensity has occurred. This property of first-order is useful to identify edges. Many methods like Roberts, Sobel, Prewitt, kirsch, and canny edge detectors compute the derivatives (row gradient and column gradient) for an entire image using a 2 × 2 or higher dimension-mask.

1.7.1.1 Gradient Operator

Gradient operator determines the changes in intensity in the adjacent pixel, and the resultant gradient magnitude emphasizes that pixels have significant local intensity changes. Some basic methods that use gradient operators are discussed in this section. Roberts edge detection convolves two 2 × 2 masks or operators (shown in Fig. 1.18) on an image in the image's x and y direction and takes less time to measure the gradient $G_x(i,j)$. Image pixels $I(i,j)$ are input to operators, and output pixels $G(i,j)$ are a magnitude of the gradient. In this method, the masks or operators are designed to produce the maximum magnitude of gradients for diagonal edges in an image.

$$G_x(i,j) = I(i,j) * (E_9 - E_5) \tag{1.25}$$

$$G_y(i,j) = I(i,j) * (E_8 - E_6) \tag{1.26}$$

$$G(i,j) = |G_x(i,j)| + |G_y(i,j)| \tag{1.27}$$

It has few problems, such as it is very sensitive to noise because of small-size operators, and it produces a weak response for blur edges. Sobel, Prewitt, and Robinson Edge detection methods use comparatively large-size operators, i.e., 3 × 3. The gradient is measured by convolving the orthogonal operators of Sobel and Prewitt at the image. Robinson Edge detection is like Sobel edge detection, but Robinson has eight sets of 3 × 3 operators. The operators' name on compass direction is also known as the Robinson compass edge detection method. Its eight operators convolve on an image that would produce a strong result if any change of intensity is observed. Compared to other convolution edge detectors, it produces an accurate gradient but takes a longer time to complete the edge detection process. All these operators or masks with their values are depicted in Fig. 1.19.

Not all non-zero gradient values of points at the edge (e.g., ramp edge) are marked as edge points. Therefore, a threshold is applied to mark points as edge

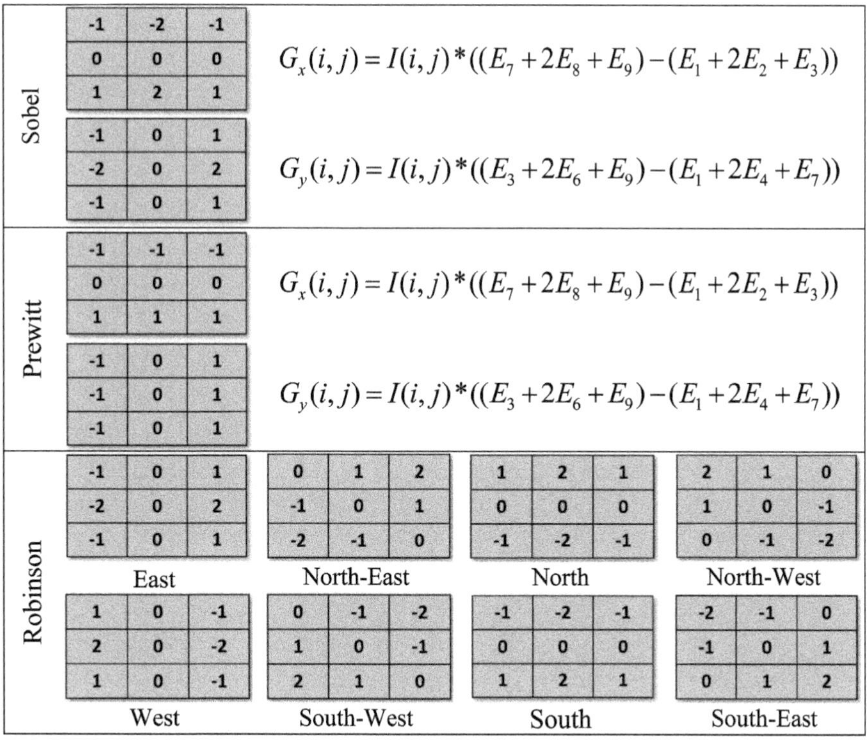

Sobel

-1	-2	-1
0	0	0
1	2	1

$$G_x(i,j) = I(i,j)*((E_7 + 2E_8 + E_9) - (E_1 + 2E_2 + E_3))$$

-1	0	1
-2	0	2
-1	0	1

$$G_y(i,j) = I(i,j)*((E_3 + 2E_6 + E_9) - (E_1 + 2E_4 + E_7))$$

Prewitt

-1	-1	-1
0	0	0
1	1	1

$$G_x(i,j) = I(i,j)*((E_7 + 2E_8 + E_9) - (E_1 + 2E_2 + E_3))$$

-1	0	1
-1	0	1
-1	0	1

$$G_y(i,j) = I(i,j)*((E_3 + 2E_6 + E_9) - (E_1 + 2E_4 + E_7))$$

Robinson

East

-1	0	1
-2	0	2
-1	0	1

North-East

0	1	2
-1	0	1
-2	-1	0

North

1	2	1
0	0	0
-1	-2	-1

North-West

2	1	0
1	0	-1
0	-1	-2

West

1	0	-1
2	0	-2
1	0	-1

South-West

0	-1	-2
1	0	-1
2	1	0

South

-1	-2	-1
0	0	0
1	2	1

South-East

-2	-1	0
-1	0	1
0	1	2

Fig. 1.19 Gradient value of all elements of the Sobel, Prewitt, and Robinson masks at different directions

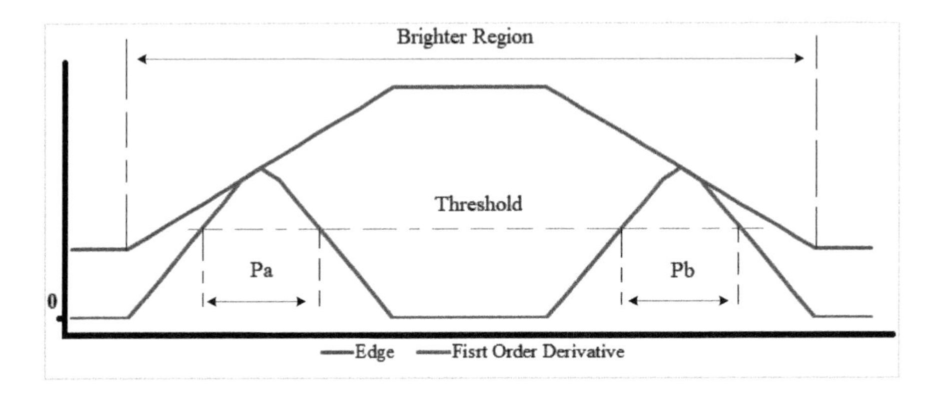

Fig. 1.20 Threshold at the absolute value of the first-order derivative (i.e., gradient value)

points if their gradient value is greater than the threshold value. The process of marking edge points in gradient-based methods by thresholding causes a thick edge detection. Refer to Fig. 1.20 for illustration, where the edges points Pa and Pb are observed at two intensity transition states of the brighter region after thresholding. These large number points cause a thick edge detection.

1.7.2 Second-Order Derivative Edge Detection

A second-order spatial derivative of increasing and decreasing intensity values produces an extreme positive value and extreme negative value, respectively (Fig. 1.21). Other than this, the second-order spatial derivative is 0 when there is linear intensity change or no intensity change. The constant or linear intensity change pixels between the dark and brighter pixels of the edge cause strong zero-crossing in the edge's second-order spatial function. The zero-crossing is very weak for the step or unit edges that can be solved using some threshold. Therefore, this zero-crossing property of second-order is beneficial to identify the mid-point of a thick edge. Laplacian is one type of second-order derivative edge detection method.

Laplacian methods use filters, for example, 2D Gaussian kernel on an image, in advance to remove unwanted noise, such as salt and paper noise, impulse noise, and Gaussian noise.

$$g(x,y) = \frac{1}{2\pi\sigma^2} e^{\frac{-(x^2+y^2)}{2\sigma^2}} \tag{1.28}$$

To generate 3 × 3 or 5 × 5 kernel of 2D Gaussian for noise filtering, the following formula is used.

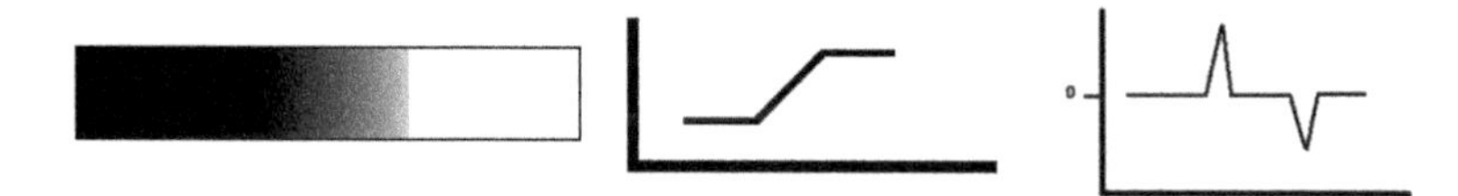

Fig. 1.21 Intensity graph of the second-order derivative of ramp edge

0	-1	0
-1	4	-1
0	-1	0

$$\nabla^2 I = 4E_5 - (E_2 + E_4 + E_6 + E_8)$$

-1	-1	-1
-1	8	-1
-1	-1	-1

$$\nabla^2 I = 8E_5 - (E_1 + E_2 + E_3 + E_4 + E_6 + E_7 + E_8 + E_9)$$

Fig. 1.22 Value of all elements of the 3 × 3 mask for Laplacian edge method

$$g(x,y) = e^{\frac{-(x^2+y^2)}{2\sigma^2}} \tag{1.29}$$

Laplacian of an image function $I(x, y)$ in the continuous domain is defined as,

$$\nabla^2 I = \frac{\partial^2 I}{\partial x^2} + \frac{\partial^2 I}{\partial y^2} \tag{1.30}$$

As shown in Fig. 1.22, there are mainly two 3 × 3 masks with either four or eight neighbor Laplacian. For detection, the Laplacian edge method uses one alone to detect the edge. The mask with four neighbor Laplacian focuses on vertical and horizontal directions, whereas the other mask focuses on all directions, including the diagonals. Laplacian of Gaussian (LoG) based methods use the Gaussian filter in computing the zero-crossing detection step.

$$LoG = \nabla^2 S = \nabla^2 (g \times I) = (\nabla^2 g) \times I \tag{1.31}$$

$$\nabla^2 g = -\frac{1}{\sqrt{2\pi\sigma^3}}\left(2 - \frac{x^2 + y^2}{\sigma^2}\right) e^{-\frac{x^2+y^2}{2\sigma^2}} \tag{1.32}$$

The 3 × 3 or 5 × 5 kernel of LoG is generated by using the following formula.

$$g(x,y) = \left(2 - \frac{x^2 + y^2}{\sigma^2}\right) e^{-\frac{x^2+y^2}{2\sigma^2}} \tag{1.33}$$

There may be very weak zero-crossing detected by applying a second-order edge detector. This difficult set a threshold to mark sharp edges if the weak zero-crossing slope is observed. (Fig. 1.23), where the strong zero-crossing *Za* and week zero-crossing *Zb* are observed. The mid-point *Ma* of *Za* is marked as an edge point. However, the mid-point calculation is become very difficult for weak zero-crossing (*Zb*) because of no points having zero value of the second-order derivative.

1.7.3 Optimal Canny Edge Detection

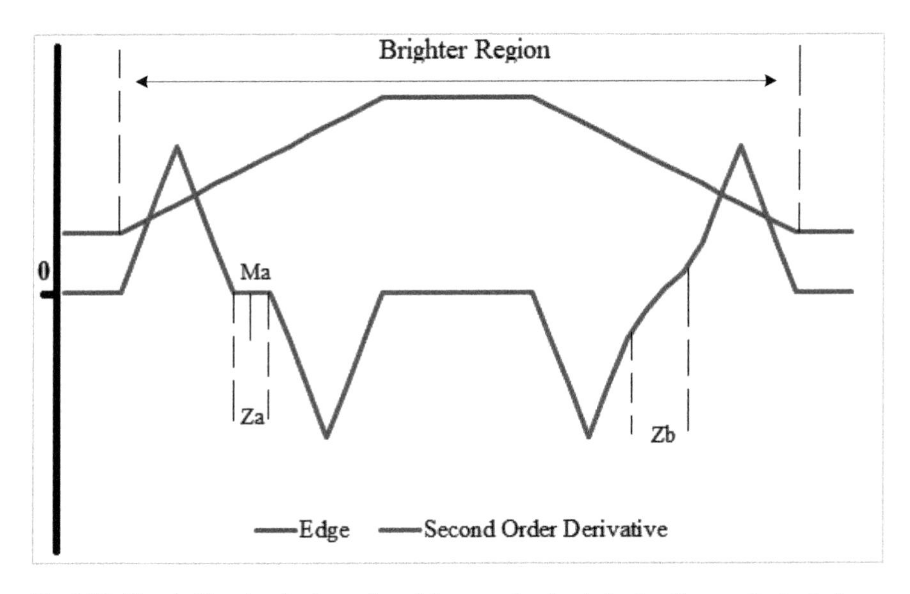

Fig. 1.23 Threshold at the absolute value of the second-order derivative (i.e., gradient value)

Unlike the first-order gradient techniques, the Canny edge detection technique utilizes gradient orientation along with gradient magnitude. The gradient is very sensitive to noise; therefore, a Gaussian filter is applied before the gradient analysis to remove noise. Canny edge measures the gradient and its direction by applying the Sobel orthogonal operators for edge detection. Next, a non-maximum suppression process is applied to detected edges. It locates an optimal edge by minimizing the distance between the detected edge and the real edge. The real edge points may locate at their neighboring pixels along the direction of the detected edge. The direction of the edge is at 90° concerning the gradient direction. The gradient magnitude is analyzed at neighboring points of a detected edge in the gradient direction. A point with the greatest gradient magnitude is assigned as a real edge point. This analysis is performed from one end of the detected edge to the other. This process ensures thin edge detection. An example is shown in Fig. 1.24; the edge in green located at the maximum gradient is set as a real edge after applying a non-maximum suppression process.

All real edges should not be true edges; therefore, the real edges are reanalyzed using double threshold methods called hysteresis thresholding. As shown in Fig. 1.25, this method histograms the magnitude of all real edges and uses two thresholds, high and low thresholds, to group them into true and false edges. For true edges, the following condition should be fulfilled.

(a) The gradient magnitude of real edges is greater than the high threshold.
(b) If the gradient magnitude of edges is less than the high threshold but greater than the low threshold, such edges' connectivity with high gradient magnitude confirms its true edges rank.

1.7.4 Edge Linkers for Boundary Detection

Sharp edges detected from canny edge techniques are not enough to segment regions in an image, but these edges help extract the boundary of regions present in an image. There are two major techniques to link the detected edges for generating boundary regions present in an image, such as local edge linkers and global edge linkers. In local edge linkers, a filter is used to analyze image pixels' properties in a defined local neighborhood. Eqs. 1.34 and 1.35 measure a similarity index for the pixels between two edges. These pixels are assigned as edge pixels if they are similar at some threshold in gradient and angle to the adjacent edges' pixels.

$$\left|\nabla I\left(x_0,y_0\right)-\nabla I(x_1,y_1)\right|\leq\left|T_g\right| \tag{1.34}$$

$$\left|\alpha\left(x_0,y_0\right)-\alpha(x_1,y_1)\right|\leq\left|T_d\right| \tag{1.35}$$

Hough transformation is a robust strategy of global edge linkers that extracts the perpendicular and parallel lines/edges from an entire image (Cui et al., 2012). It is

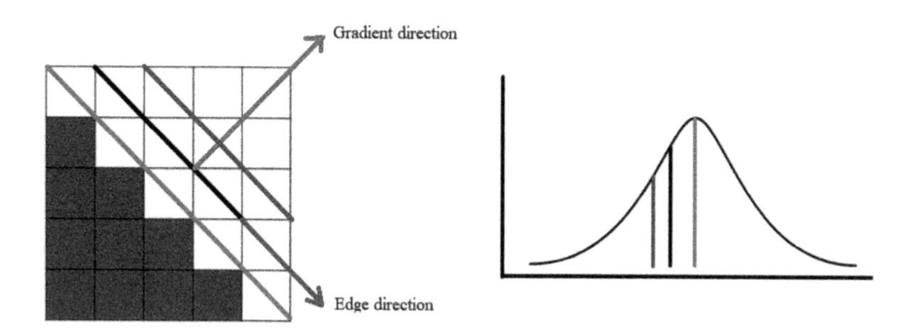

Fig. 1.24 A non-maximum suppression process to detect real edges. Here, the detected edge is highlighted in black, and possible real edges are highlighted in green and purple

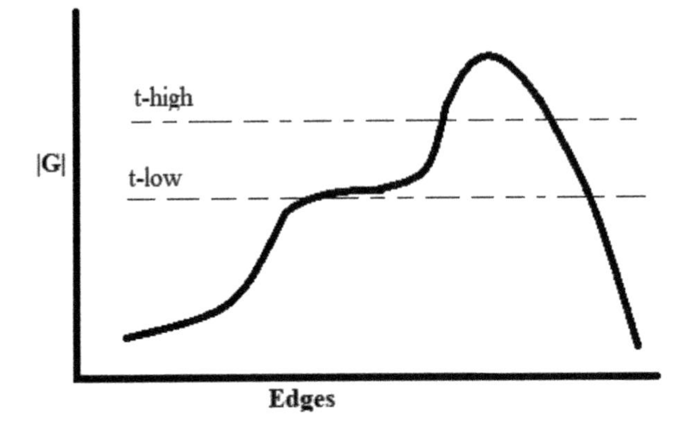

Fig. 1.25 Hysteresis thresholding method, wherein two thresholds are used to group true and false edges

a linear transform method that is less effective than noise. It considers the line characteristics in image space, i.e., (slope) and (intersection) instead of line points (x_2, y_2). Hough transformation transforms the image space into parameter space m and b. However, the vertical line makes a problem, that is, unbound values are obtained on parameter space. Therefore, m and b are replaced with the r (distance from the line to a particular pixel or point) and θ (angle of the vector from the original pixel or point to the line). Figure 1.26 describes these parameters where the line is defined as:

$$y = \left(-\frac{\cos\theta}{\sin\theta} \right) x + \left(\frac{r}{\sin\theta} \right) \tag{1.36}$$

where r is derived as:

$$r(\theta) = x\cos\theta + y\sin\theta \tag{1.37}$$

This method stores the parameters for each line in a matrix called an accumulator (parameter space). One dimension of the matrix is the distance, and the other is the angle. So, all the lines with the same value of parameters r and θ is examined, and the line with the highest number of pixels or points is selected to represent the image edges. Moreover, the parameter space is used to find the parallel and perpendicular lines. As it is known from Eq. 1.36, the line is sinusoidal, and the peaks of parallel lines image space are aligned vertically in parameter space, and peaks of perpendicular lines in image space are about $\neq/2$ in parameter space. By using the Hough transformation, all parallel and perpendicular edges are extracted. In addition, the ratio of gray value between the two sides of edges is measured, and if it is less than the predefined threshold, then the edge is removed (Cui et al., 2012).

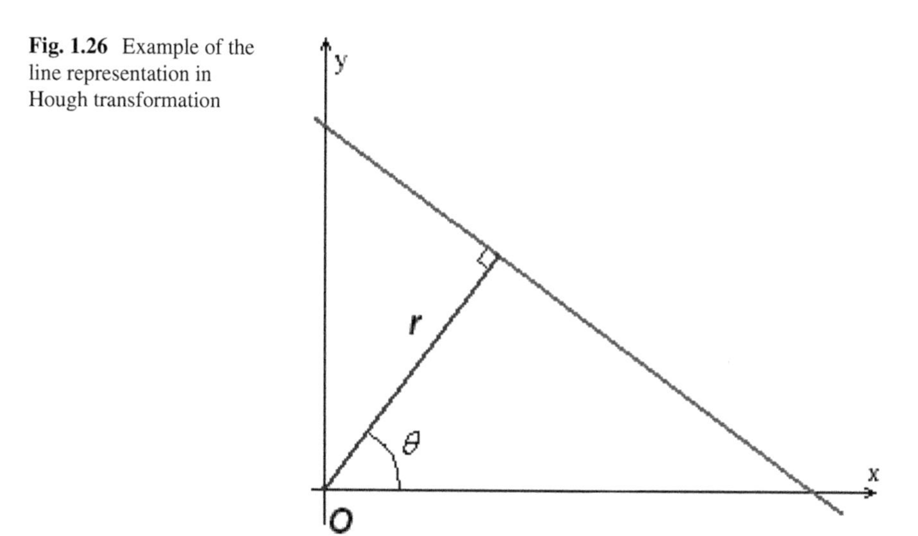

Fig. 1.26 Example of the line representation in Hough transformation

References

Abad, F., Garcia-Consuegra, J., & Cisneros, G. (2000). Merging regions based on the VDM distance. In IGARSS 2000. IEEE 2000 International Geoscience and Remote Sensing Symposium. Taking the Pulse of the Planet: The Role of Remote Sensing in Managing the Environment. Proceedings (Cat. No. 00CH37120) (Vol. 2, pp. 615–617). IEEE.

Almeida, J., Barbosa, L., Pais, A., & Formosinho, S. (2007). Improving hierarchical cluster analysis: A new method with outlier detection and automatic clustering. *Chemometrics and Intelligent Laboratory Systems, 87*(2), 208–217.

Anil, P. N., & Natarajan, S. (2010). Automatic road extraction from high resolution imagery based on statistical region merging and skeletonization. *International Journal of Engineering Science and Technology, 2*(3), 165–171.

Bernsen, J. (1986). *Dynamic thresholding of gray-level images.* Paper presented at the Proc. Eighth Int'l conf. Pattern Recognition, Paris, 1986.

Bezdek, J. C. (1980). A convergence theorem for the fuzzy ISODATA clustering algorithms. *IEEE Transactions on Pattern Analysis and Machine Intelligence, 2*(1), 1–8.

Bins, L. S. A., Fonseca, L. G., Erthal, G. J., & Ii, F. M. (1996). Satellite imagery segmentation: a region growing approach. *Simpósio Brasileiro de Sensoriamento Remoto, 8*(1996), 677–680.

Boukharouba, S., Rebordao, J. M., & Wendel, P. L. (1985). An amplitude segmentation method based on the distribution function of an image. *Computer vision, graphics, and image processing, 29*(1), 47–59. https://doi.org/10.1016/s0734-189x(85)90150-1

Chih-Cheng, H., Kulkarni, S., & Bor-Chen, K. (2011). A new weighted fuzzy c-means clustering algorithm for remotely sensed image classification. *IEEE Journal of Selected Topics in Signal Processing, 5*(3), 543–553. https://doi.org/10.1109/jstsp.2010.2096797

Cho, S., Haralick, R., & Yi, S. (1989). Improvement of Kittler and Illingworth's minimum error thresholding. *Pattern recognition, 22*(5), 609–617.

Chuang, K.-S., Tzeng, H.-L., Chen, S., Wu, J., & Chen, T.-J. (2006). Fuzzy c-means clustering with spatial information for image segmentation. *Computerized Medical Imaging and Graphics, 30*(1), 9–15. https://doi.org/10.1016/j.compmedimag.2005.10.001

Cui, S., Yan, Q., & Reinartz, P. (2012). Complex building description and extraction based on Hough transformation and cycle detection. *Remote Sensing Letters, 3*(2), 151–159.

Dare, P. M. (2005). Shadow analysis in high-resolution satellite imagery of urban areas. *Photogrammetric Engineering & Remote Sensing, 71*(2), 169–177.

Derrien, M., & Le Gléau, H. (2007). Temporal-differencing and region-growing techniques to improve twilight low cloud detection from SEVIRI data. In Proceedings of the Joint 2007 EUMETSAT

Meteorological Satellite Conference and the 15th Satellite Meteorology and Oceanography Conference of the American Meteorological Society (Vol. 2428, p. 2428), Amsterdam: The Netherlands.
Devereux, B. J., Amable, G. S., & Posada, C. C. (2004). An efficient image segmentation algorithm for landscape analysis. International Journal of Applied Earth Observation and Geoinformation 6(1) 47-61. https://doi.org/10.1016/j.jag.2004.07.007.
Dixon, S. J., Heinrich, N., Holmboe, M., Schaefer, M. L., Reed, R. R., Trevejo, J., & Brereton, R. G. (2009). Use of cluster separation indices and the influence of outliers: Application of two new separation indices, the modified silhouette index and the overlap coefficient to simulated data and mouse urine metabolomic profiles. *Journal of Chemometrics, 23*(1), 19–31.
Doyle, W. (1962). Operations useful for similarity-invariant pattern recognition. *Journal of the ACM (JACM), 9*(2), 259–267.
Du, Q., & Gunzburger, M. (2002). Grid generation and optimization based on centroidal Voronoi tessellations. *Applied Mathematics and Computation, 133*(2), 591–607.
Faruquzzaman, A. B. M., Paiker, N. R., Arafat, J., & Ali, M. A. (2008). A survey report on image segmentation based on split and merge algorithm. *IETECH Journal of Advanced Computations, 2*(2), 86–101.
Faruquzzaman, A. B. M., Paiker, N. R., Arafat, J., Ali, M. A., & Sorwar, G. (2009). Robust Object Segmentation using Split-and-Merge. *International Journal of Signal and Imaging Systems Engineering, 2*(1/2), 70. https://doi.org/10.1504/IJSISE.2009.029332
Gonzalez, R. C., Woods, R. E., & Eddins, S. L. (2003). Digital Image Processing Using MATLAB.
Guha, S., Rastogi, R., & Shim, K. (2000). ROCK: A robust clustering algorithm for categorical attributes. *Information Systems, 25*(5), 345–366.
Guha, S., Rastogi, R., & Shim, K. (2001). Cure: An efficient clustering algorithm for large databases. *Information Systems, 26*(1), 35–58.
Hamerly, G., & Elkan, C. (2002). *Alternatives to the k-means algorithm that find better clusterings.* Paper presented at the Proceedings of the Eleventh International Conference on Information and Knowledge Management, McLean, Virginia, USA.
Hathaway, R. J., Bezdek, J. C., & Hu, Y. (2000). Generalized fuzzy c-means clustering strategies using Lp norm distances. *IEEE Transactions on Fuzzy Systems, 8*(5), 576–582.
Hemachander, S., Verma, A., Arora, S., & Panigrahi, P. K. (2007). Locally adaptive block thresholding method with continuity constraint. *Pattern Recognition Letters, 28*(1), 119–124.
Horowitz, S. L., & Pavlidis, T. (1976). Picture Segmentation by a Tree Traversal Algorithm. *Journal of the ACM, 23*(2), 368–388. https://doi.org/10.1145/321941.321956
Huang, L. K., & Wang, M. J. J. (1995). Image thresholding by minimizing the measures of fuzziness. *Pattern recognition, 28*(1), 41–51.
Hung, W. L., Yang, M. S., & Chen, D. H. (2008). Bootstrapping approach to feature-weight selection in fuzzy *c*-means algorithms with an application in color image segmentation. *Pattern Recognition Letters, 29*(9), 1317–1325.
Huynh Van, L., & Jong-Myon, K. (2009, August 20–24). *A generalized spatial fuzzy c-means algorithm for medical image segmentation.* Paper presented at the Fuzzy Systems, 2009. IEEE International Conference on FUZZ-IEEE 2009.
Isa, N. A. M., Salamah, S. A., & Ngah, U. K. (2009). Adaptive fuzzy moving K-means clustering algorithm for image segmentation. *IEEE Transactions on Consumer Electronics, 55*(4), 2145–2153.
Ismail, S. M., Abdullah, S. N. H. S., & Fauzi, F. (2018). Statistical binarization techniques for document image analysis. *Journal of Computer Science, 14*(1), 23–36.
Jawahar, C., Biswas, P., & Ray, A. (1997). Investigations on fuzzy thresholding based on fuzzy clustering. *Pattern Recognition, 30*(10), 1605–1613.
Jianzhuang, L., Wenqing, L., & Yupeng, T. (1991). *Automatic thresholding of gray-level pictures using two-dimension Otsu method.* Paper presented at the Circuits and Systems, 1991. 1991 International Conference on Conference Proceedings, China.
Jin, Z., Lou, Z., Yang, J., & Sun, Q. (2007). Face detection using template matching and skin-color information. *Neurocomputing, 70*(4–6), 794–800. https//doi.org/10.1016/j.neucom.2006.10.043
Jinhai, C., & Zhi-Qiang, L. (1998, August 16–20). *A new thresholding algorithm based on all-pole model.* Paper presented at the Pattern Recognition, 1998. Fourteenth International Conference on Proceedings.

Kampke, T., & Kober, R. (1998, August 16–20). *Nonparametric optimal binarization.* Paper presented at the Pattern Recognition, 1998. Fourteenth International Conference on Proceedings.
Karypis, G., Han, E. H., & Kumar, V. (1999). Chameleon: Hierarchical clustering using dynamic modeling. *Computer, 32*(8), 68–75.
Kersten, P. R. (1999). Fuzzy order statistics and their application to fuzzy clustering. *IEEE Transactions on Fuzzy Systems, 7*(6), 708–712.
Khoshelham, K., Li, Z., & King, B. (2005). A Split-and-Merge Technique for Automated Reconstruction of Roof Planes. *Photogrammetric Engineering & Remote Sensing, 71*(7), 855–862. https://doi.org/10.14358/PERS.71.7.855
Khotanzad, A., & Bouarfa, A. (1990). Image segmentation by a parallel, non-parametric histogram based clustering algorithm. *Pattern Recognition, 23*(9), 961–973.
Kittler, J., & Illingworth, J. (1986). Minimum error thresholding. *Pattern Recognition, 19*(1), 41–47.
Lang, X., Zhu, F., Hao, Y., & Ou, J. (2008). *Integral image based fast algorithm for two-dimensional Otsu thresholding.* Paper presented at the Image and Signal Processing, 2008. Congress on CISP'08.
Lloyd, D. (1985). Automatic target classification using moment invariant of image shapes. *IDN AW126, RAE, Farnborough, Reino Unido.*
Lucieer, A., & Stein, A. (2002). Existential uncertainty of spatial objects segmented from satellite sensor imagery. *IEEE Transactions on Geoscience and Remote Sensing, 40*(11), 2518–2521. https://doi.org/10.1109/TGRS.2002.805072
Manousakas, I. N., Undrill, P. E., Cameron, G. G., & Redpath, T. W. (1998). Split-and-Merge Segmentation of Magnetic Resonance Medical Images: Performance Evaluation and Extension to Three Dimensions. *Computers and Biomedical Research, 31*(6), 393–412. https://doi.org/10.1006/cbmr.1998.1489
Mashor, M. Y. (2000). Hybrid training algorithm for RBF network. *International Journal of The Computer, The Internet and Management, 8*(2), 50–65.
Mazonakis, M., Damilakis, J., Varveris, H., Prassopoulos, P., Gourtsoyiannis, N. (2001). Image segmentation in treatment planning for prostate cancer using the region growing technique. *The British Journal of Radiology, 74*(879), 243–249. https/doi.org/10.1259/bjr.74.879.740243
Meethongjan, K., Dzulkifli, M., Rehman, A., & Saba, T. (2010). Face recognition based on fusion of Voronoi diagram automatic facial and wavelet moment invariants. *International Journal Video Process Image Process Netw Secur, 10*(4), 1–8.
Murthy, C. A., & Pal, S. K. (1990). Fuzzy thresholding: Mathematical framework, bound functions and weighted moving average technique. *Pattern Recognition Letters, 11*(3), 197–206.
Nedzved, A., Ablameyko, S., & Pitas, I. (2000). Morphological segmentation of histology cell images. In, 2000. Published by the IEEE Computer Society.
Olivo, J. C. (1994). Automatic threshold selection using the wavelet transform. *CVGIP: Graphical Models and Image Processing, 56*(3), 205–218.
Olson, C. F. (1995). Parallel algorithms for hierarchical clustering. *Parallel Computing, 21*(8), 1313–1325.
Otsu, N. (1975). A threshold selection method from gray-level histograms. *Automatica, 11*(285-296), 23–27.
Pal, S. K., & Rosenfeld, A. (1988). Image enhancement and thresholding by optimization of fuzzy compactness. *Pattern Recognition Letters, 7*(2), 77–86.
Pavlidis, T., & Horowitz, S. L. (1974) Segmentation of Plane Curves. *IEEE Transactions on Computers C-23*(8), 860–870. https//doi.org/10.1109/T-C.1974.224041
Pham, D. L., Xu, C., & Prince, J. L. (2000). Current methods in medical image segmentation 1. *Annual Review of Biomedical Engineering, 2*(1), 315–337.
Pohle, R., & Toennies, K. D. (2001). A new approach for model-based adaptive region growing in medical image analysis. In International conference on computer analysis of images and patterns (pp. 238–246). Springer, Berlin, Heidelberg.
Pun, T. (1981). Entropic thresholding, a new approach. *Computer Graphics and Image Processing, 16*(3), 210–239.
Ramesh, N., Yoo, J. H., & Sethi, I. K. (1995). Thresholding based on histogram approximation. *IEE Proceedings - Vision, Image and Signal Processing, 142*(5), 271–279. https://doi.org/10.1049/ip-vis:19952007

Rau, J.-Y., & Chen, L.-C. (2003). Robust Reconstruction of Building Models from Three-Dimensional Line Segments. *Photogrammetric Engineering & Remote Sensing, 69*(2), 181–188. https://doi.org/10.14358/PERS.69.2.181

Ridler, T., & Calvard, S. (1978). Picture thresholding using an iterative selection method. *IEEE Transactions on Systems, Man and Cybernetics, 8*(8), 630–632.

Rosenfeld, A., & De La Torre, P. (1983). Histogram concavity analysis as an aid in threshold selection. *IEEE Transactions on Systems, Man and Cybernetics, SMC-13*(2), 231–235. https://doi.org/10.1109/tsmc.1983.6313118

Sahoo, P. K., Soltani, S., & Wong, A. K. (1988). A survey of thresholding techniques. *Computer Vision, Graphics, and Image Processing, 41*(2), 233–260.

Samopa, F., & Asano, A. (2009). Hybrid image thresholding method using edge detection. *International Journal of Computer Science and Network Security, 9*(4), 292–299.

Sengupta, K., Shiqin, W., Ko, C. C., & Burman, P. (2000). Automatic face modeling from monocular image sequences using modified non parametric regression and an affine camera model. In Proceedings Fourth IEEE International Conference on Automatic Face and Gesture Recognition (Cat. No. PR00580) (pp. 524–529).

Sezgin, M. (2004). Survey over image thresholding techniques and quantitative performance evaluation. *Journal of Electronic Imaging, 13*(1), 146–168.

Shih, F. Y. (2010). *Image processing and pattern recognition: Fundamentals and techniques.* Wiley.

Siddiqui, F. U. (2012). Enhanced clustering algorithms for gray-scale image segmentation. Master dissertation, Universiti Sains Malaysia.

Siddiqui, F. U., & Mat Isa, N. A. (2011). Enhanced moving K-means (EMKM) algorithm for image segmentation. *IEEE Transactions on Consumer Electronics, 57*(2), 833–841. https://doi.org/10.1109/tce.2011.5955230

Sun, Y., & Bhanu, B. (2009). Symmetry integrated region-based image segmentation. In 2009 IEEE Conference on Computer Vision and Pattern Recognition, pp. 826–831.

Taghizadeh, M., & Hajipoor, M. (2011). *A hybrid algorithm for segmentation of MRI images based on edge detection.* Paper presented at the Soft Computing and Pattern Recognition (SoCPaR), 2011 international conference of.

Tizhoosh, H. R. (2005). Image thresholding using type II fuzzy sets. *Pattern Recognition, 38*(12), 2363–2372.

Trussell, H. (1979). Comments on. *IEEE Transactions on Systems, Man and Cybernetics, 9*(5), 311–311.

Tsai, D. M. (1995). A fast thresholding selection procedure for multimodal and unimodal histograms. *Pattern Recognition Letters, 16*(6), 653–666.

Vantaram, S. R., & Saber, E. (2012). Survey of contemporary trends in color image segmentation. *Journal of Electronic Imaging*, 040901-040901-040901-040928.

Viola, P., & Jones, M. J. (2004). Robust real-time face detection. *International Journal of Computer Vision, 57*(2), 137–154.

Wang, X., Wang, Y., & Wang, L. (2004). Improving fuzzy c-means clustering based on feature-weight learning. *Pattern Recognition Letters, 25*(10), 1123–1132.

Whatmough, R. (1991). Automatic threshold selection from a histogram using the "exponential hull". *CVGIP: Graphical Models and Image Processing, 53*(6), 592–600.

Xiao, Y., Cao, Z., & Zhuo, W. (2011). Type-2 fuzzy thresholding using GLSC histogram of human visual nonlinearity characteristics. *Optics Express, 19*(11), 10656–10672.

Xu, R., & Wunsch, D. (2005). Survey of clustering algorithms. *IEEE Transactions on Neural Networks, 16*(3), 645–678.

Yang, M. S., Hwang, P. Y., & Chen, D. H. (2004). Fuzzy clustering algorithms for mixed feature variables. *Fuzzy Sets and Systems, 141*(2), 301–317.

Yanni, M., & Horne, E. (1994). *A new approach to dynamic thresholding.* Paper presented at the EUSIPCO'94: 9th European Conf. Sig. Process.

Yong, Y., Chongxun, Z., & Pan, L. (2004, September 14–16). *Image thresholding based on spatially weighted fuzzy c-means clustering.* Paper presented at the Computer and Information Technology, 2004. The fourth international conference on CIT '04.

Zhang, T., Ramakrishnan, R., & Livny, M. (1997). BIRCH: A new data clustering algorithm and its applications. *Data Mining and Knowledge Discovery, 1*(2), 141–182.

Zhao, Y., Karypis, G., & Fayyad, U. (2005). Hierarchical clustering algorithms for document datasets. *Data Mining and Knowledge Discovery, 10*(2), 141–168.

Chapter 2
Partitioning Clustering Techniques

2.1 Partitioning Clustering for Image Segmentation

Partitioning clustering is an unsupervised technique to segment a given image. The hard partitioning clustering and fuzzy partitioning clustering techniques are two main types of partitioning clustering, as mentioned in Chap. 1. However, both have limitations like they converge to a poor local minimum location. Implementation and limitations of major techniques of hard and fuzzy partitioning clustering are discussed in detail in this chapter.

2.2 k-means Clustering

The k-means technique is the renowned clustering technique (MacQueen, 1967). It employs the hard membership function for segmenting pixels, and the hard membership restricts each pixel to become a member of only one cluster. First, the k-means clustering initializes the cluster's center value called the centroid. Next, all pixels are assigned to the nearest cluster based on minimum Euclidean distance. Mean pixels' value inside their related clusters is measured to update all clusters' centroid values. Pixels assigning and centroid calculation is an iterative process and is continued until no significant changes are obtained in the clusters' centroid value. At the end of this iterative process, the solution should converge to the global optimum solution.

F.U. Siddiqui, A. Yahya, *Clustering Techniques for Image Segmentation*,
https://doi.org/10.1007/978-3-030-81230-0_2

2.2.1 Implementation of k-means Clustering

Consider an image *I*with the size of $N \times M$ (here, N and M represent several rows and columns, respectively) to be clustered into k several clusters. Let, $p_i(x, y)$ be the ith pixels and c_j be the jth centroid, where $i = 1, 2, 3, \ldots(N \times M)$ and $j = 1, 2, 3, \ldots k$. According to the study (Weisstein, 2004), k-means clustering is implemented as follows:

input: test image with the size of $N \times M$
input: k number of clusters
input: generate a random value of the centroids (c_i)
output: centroids value and segmented image

initialize
// generate a segmented image with the test image's size, and it contains no information on their pixels.
while
  for $i \leftarrow 1$ **to** N **do**
    for $j \leftarrow 1$ **to** M **do**
      // measure the distance between all $p_i(x, y)$ of the test images and centroids.
      // $p_i(x, y)$ with minimum distance assigned to their centroids. In other words, labeling the segmented image pixels with the centroid value of their nearest clusters.
      // measure the (n_j) number pixels in each cluster.
    end for
  end for
  //calculate the new value of centroids using Eq. 2.1.

$$c_j = \frac{1}{n_j} \sum_{i \in c_j} p_i(x, y) \qquad \textbf{(2.1)}$$

  if $c_j \neq c_i$
    // erase all information from the segmented image.
    // $c_i = c_j$, erase all information from c_j.
    // Repeat both for loops.
  else
    // return the segmented image and information centroid (c_j)
  end if
end while

For better illustration, the flow diagram of the k-means clustering is depicted in Fig. 2.1.

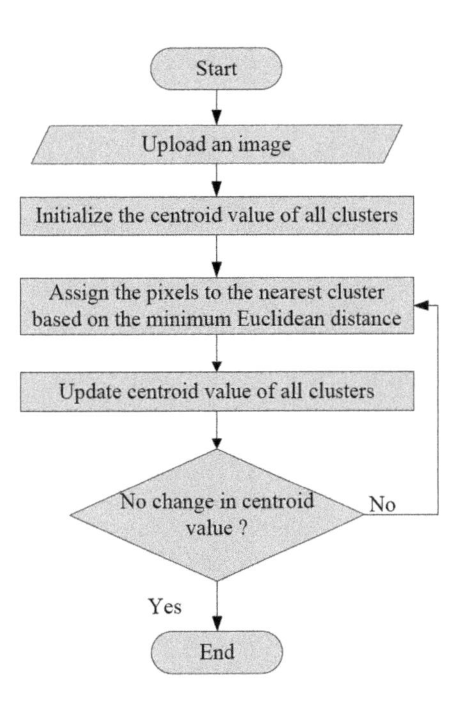

Fig. 2.1 Flow diagram of k-means clustering

2.2.2 *Limitations of k-means Clustering*

Generally, the k-means are not robust to the outliers and initialization settings. It is very normal with k-means that different initial clusters' centroid locations could yield empty clusters and other final solutions (Bradley & Fayyad, 1998). In most cases, the final solution is converged to a poor local minimum location (Kaufman & Rousseeuw, 2009) (Ester et al., 1995) (Sulaiman & Isa, 2010). Kaufman and Rousseeuw (2009) consider that the cluster's labeling by the cluster pixels' mean value is liable for the empty clusters problem. In other words, the over sensitiveness of k-means centroid to the outliers could result in the nearest cluster becoming an empty cluster or a dead centroid. Thus, the final solution may converge to the poor local minima. To overcome outlier sensitivity, the Partitioning Around Medoids (PAM) was proposed by Kaufman and Rousseeuw (1990). A medoid is defined as a representative member of a cluster with a dataset whose average dissimilarity to all the members in the cluster is minimal (Kaufman & Rousseeuw, 2009). This study concludes that a cluster's labeling using medoid value is far more effective than the mean value to reduce outlier sensitivity. This solution is outperformed in the k-means clustering on smaller data but does not work well if the data is large (Barioni et al., 2008).

Furthermore, PAM's computational complexity is very high compared to k-means, and it's also unable to solve the initialization problem of k-means that encourages the empty cluster hitch. Another variation of k-means called k-medians clustering was proposed to produce better segmentation results (Anderson et al., 2006). It employed the median to measure the centroid of all clusters. The main objective of the k-medians is to minimize the mean absolute deviation that makes it less sensitive to outliers compared to the k-means and k-medoid. If k-median is compared to k-means, the k-median is more sensitive to initialization parameters and often generates empty clusters. As a result, the final solution of k-median may converge to an optimum local location. It is concluded from the discussion that the main problems of k-means are empty clusters and trapped centroids at local minima, and both problems are interrelated to each other and may occur if points are assigned incorrectly.

2.2.3 Illustration of k-means Clustering's Limitations

As mentioned above, the k-means clustering assigns each pixel to the respective cluster based on minimum Euclidean distance. But there is a high chance of poor pixel assignment if the pixel has the same minimum Euclidean distance to two or more adjacent clusters (Siddiqui & Isa, 2012). Consequently, the pixel may be assigned to the higher variance cluster instead of the low variance cluster. The lower variance cluster has probably no chance to become a part of the clustering process and could be trapped at poor local minima or ended as an empty cluster. This empty cluster is also known as the dead center. Figure 2.2 visualizes the abovementioned phenomena where points A, B, and C are the centroids of the clusters adjacent to point D. The Euclidean distance between point D and its adjacent clusters A, B, and C is the same. Thus, point D may be assigned to the higher variance cluster. The lower variance cluster could easily trap into its present value with no further chance of updating or possibly being converted into an empty cluster without having a chance to be updated in the clustering process.

For a better illustration of the empty clusters and trapped centroid limitations of k-means clustering, a "*Bridge*" image from a publicly available dataset is segmented by k-means as shown in Fig. 2.3. Figure 2.3 (c) and (e) are the segmented images into three and six clusters by the k-means clustering, respectively. In addition, the histograms of the test image before and after the segmentation process with

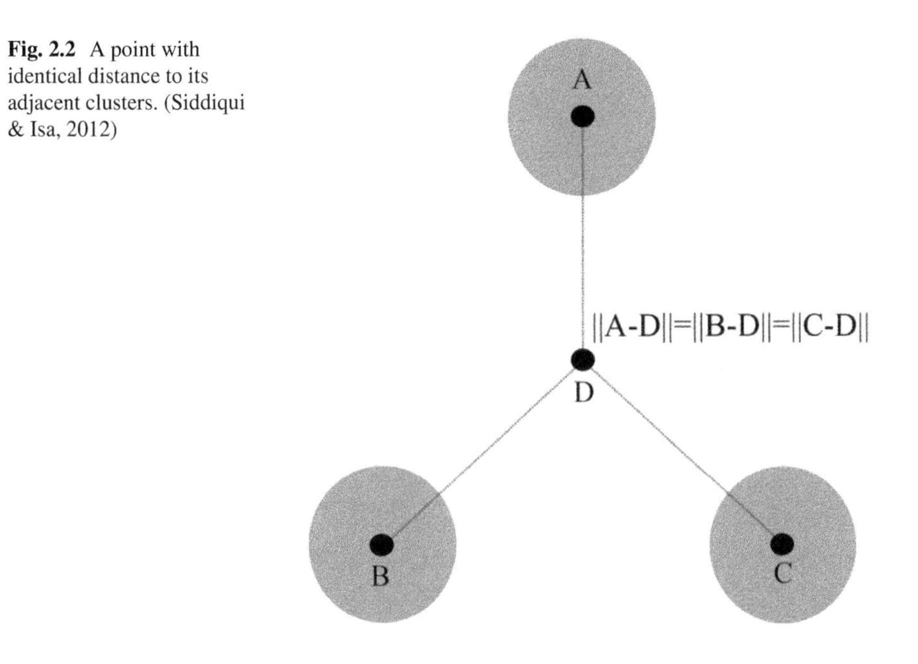

Fig. 2.2 A point with identical distance to its adjacent clusters. (Siddiqui & Isa, 2012)

different numbers of clusters (i.e., three and six) are shown in Fig. 2.3 (b), (d), and (f), respectively. Comparing histograms of the test image and k-means segmented image with three clusters confirms that k-means produced inappropriate centroids positioning where the final centroids are not located at the appropriate peak's original histogram as shown in Fig. 2.3 (d). In other words, the final centroids are trapped at the non-active region of data, and incomplete or inadequate detail from the image is segmented. For example, the bridge's ropes in Fig. 2.3 (c) are not segmented with sharp edges, but unnecessary background information like the cloud's dark shade is clustered. Whereas in Fig. 2.3 (e), the image has been segmented into five clusters, although the initial setting was six clusters. Figure 2.3 (f) also proves that the k-means clustering cannot avoid empty clusters. At the end of the clustering process, only five clusters are assigned to the "Bridge" image than the initial set value of, that is, six clusters.

Besides the qualitative analysis, the quantitative analysis is also conducted for performance assessment of k-means clustering, and the metrics like MSE, INTER, $F(I)$, $F'(I)$, and $Q(I)$ are measured for the segmented image. From the tabulated results in Table 2.1, the large value of MSE and the small value of INTER confirm the poor convergence of the k-means clustering. Whereas the large value of the $F(I)$, $F'(I)$, and $Q(I)$ shows the less homogenous segmented image because of the empty cluster.

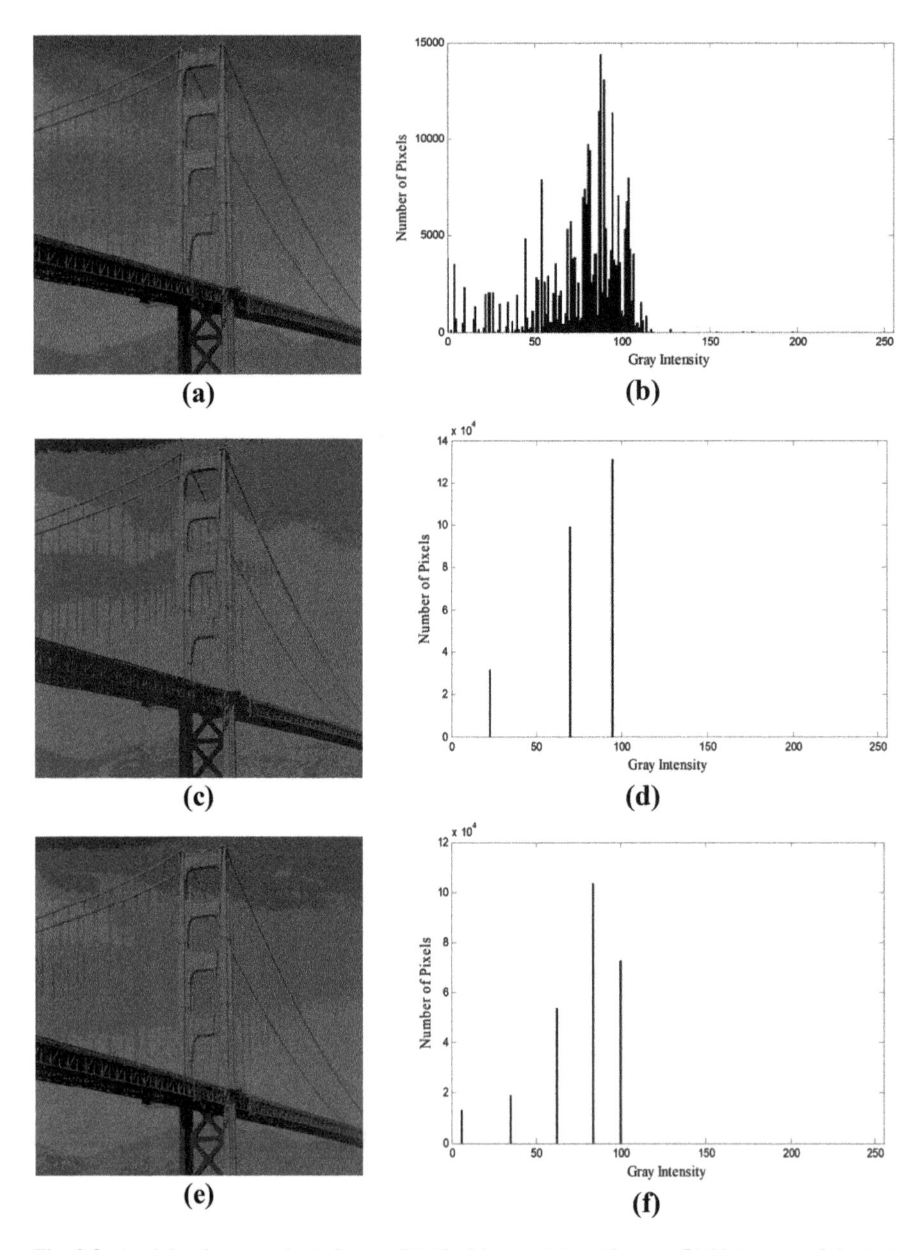

Fig. 2.3 Applying k-mean clustering on "*Bridge*" image (**a**) test image, (**b**) histogram of the test image, (**c**) segmented image at $k = 3$, (**d**) histogram of the segmented image, (**e**) segmented image at $k = 6$, and (**f**) histogram of the segmented image

Table 2.1 Quantitative analysis of k-means clustering using the "*Bridge*" image

Quantitative Test	Bridge	Bridge
Initialized number of clusters	3	6
Final number of clusters	3	5
MSE	100.942	36.20542
INTER	48	50.7
$F(I)$ 1×10^7	4.2307	4.5801
$F'(I)$	3.0765	5.7863
$Q(I)$	128.79	4251.58

2.3 Moving K-means Clustering

In 2000, the moving k-means (MKM) clustering was proposed to overcome the limitations of the k-means clustering (Mashor, 2000). The research concluded that the trapped centroid at the non-active region is the main reason for producing the dead centroid (i.e., empty cluster) and causing the solution of the k-means clustering to converge at the optimum local location (Mashor, 2000). In detail, the empty clusters are generated because of the poor initial positioning of centroids. For instance, the centroids initialized at non-active regions. Because of this phenomenon, the centroids may not be updated in the clustering process and may trap in the non-active region. MKM clustering reduces this unsuitable partitioning of the data by using some fitness criteria for clusters to keep them in an active region of the data. Based on this criterion, the fitness is continuously verified during the calculation of centroids. In any inadequate condition, the cluster with the lowest fitness value withdraws all its member pixels. It moves towards the active region by taking the cluster's member pixels with the highest fitness value. This relocation of pixels between the clusters maintains the same variance value for all clusters, and therefore the MKM becomes less sensitive to the initial settings.

2.3.1 Implementation of MKM Clustering

Based on published information on MKM clustering (Mashor, 2000), the steps of MKM clustering are as follows:

input: test image with the size of $N \times M$
input: k number of clusters
input: generate a random value of the centroids (c_i)
input: set $\alpha_a = \alpha_b = \alpha_o$. (here, α_a, α_b and α_o are the small constants. The α_o value is in the range $0 < \alpha_o < 1/3$).
output: centroids value and segmented image

initialize
// generate a segmented image with the test image's size, and it contains no information on their pixels.
while
 for $i \leftarrow 1$ **to** N **do**
  for $j \leftarrow 1$ **to** M **do**
   // measure the distance between all $p_i(x, y)$ of the test images and centroids.
   // $p_i(x, y)$ with minimum distance assigned to their centroids. In other words, labeling the segmented image pixels with the centroid value of their nearest clusters.
   // measure the (n_j) number pixels in each cluster.
  end for
 end for
 // calculate the new value of centroids using Eq. 2.1.
 // measure the fitness value for all clusters using Eq. 2.2.

$$f(c_j) = \sum_{i \in c_j} (\|p_i(x, y) - c_j\|)^2 \qquad \textbf{(2.2)}$$

 // find the cluster with the smallest and largest fitness value, i.e. c_s and c_l, respectively.
 if $f(c_s) < \alpha_a f(c_l)$ ***/ fitness verification step**
  // analyzed $p_i(x, y)$ of segmented image association with clusters
  if $p_i(x, y) < c_l$, here $i \in c_l$
   // c_s withdraw all its pixels
   // Assign those pixels of c_l to c_s and leave the rest of the pixels in c_l.

   // Recalculate the cluster's centroid position with the smallest and the largest fitness value using the following equations.

$$c_s = \frac{1}{n_{c_s}} \sum_{i \in c_s} p_i(x, y) \qquad \textbf{(2.3)}$$

$$c_l = \frac{1}{n_{c_l}} \sum_{i \in c_l} p_i(x, y) \qquad \textbf{(2.4)}$$

The flow diagram of the MKM clustering is shown in Fig. 2.4.

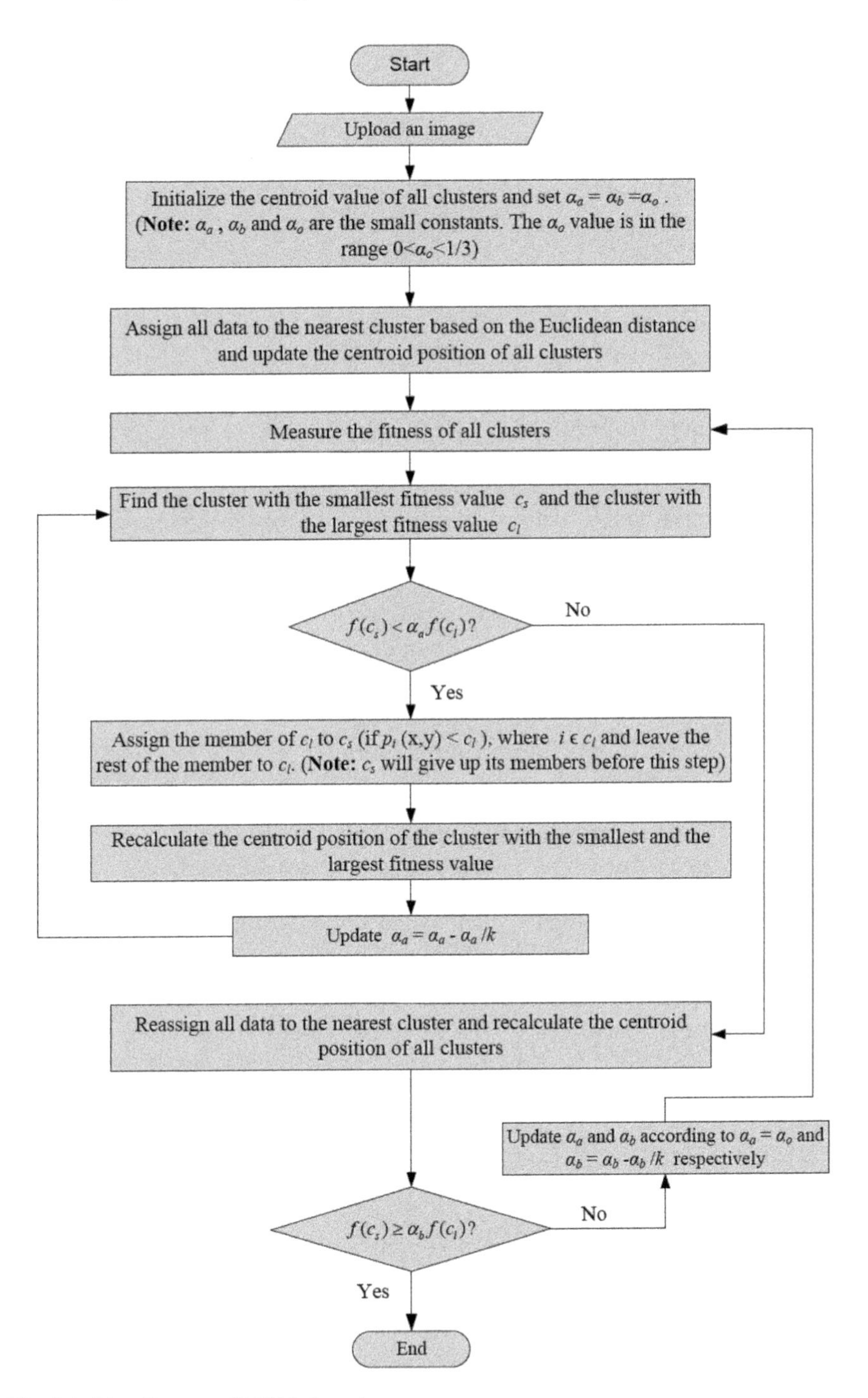

Fig. 2.4 Flow diagram of MKM clustering

2.3.2 Limitations of MKM Clustering

In MKM clustering, the cluster with the smallest fitness value (c_s) will drop its all member pixels and $c_s$move toward the active region by receiving the members, pixels with intensity value less than (i.e.,) cluster with the largest fitness value (c_l). Whereas the members of c_l having a value greater than c_l will remain a member of c_l (Fasahat Ullah Siddiqui & Isa, 2011). Based on equations 2.3 and 2.4, the c_s and c_l will be updated at the end of every iteration of the clustering process.

The transferring concept of the MKM clustering process has some major limitations, which are as follows:

- Although the fitness condition is applied to avoid centroid to trap at the non-active region, the same fitness condition causes an uncertain scenario that can probably be occurred in the clusters c_s(Fasahat Ullah Siddiqui & Isa, 2011). In detail, the cluster c_s may contain similar value members (i.e., equal to the cluster's centroid value). This causes a low Euclidean distance between the cluster's centroid and its members and produces a low fitness value for that cluster. Thus, forcing the cluster $c_s$to drop its members and become an empty cluster opposes clustering's main objective, i.e., divides data into a defined number of clusters. This increases the intra-cluster variance of clusters and produces poor segmentation.
- The transferring process of the c_l members in MKM clustering may cause poor segmentation. In the transferring process of the c_l members, the members are divided into two groups: (a) members with a value greater than the centroid value (c_l) and (b) members with a value less than the centroid value (c_l) (Siddiqui & Isa, 2011). Members of c_l the first group will remain members of that cluster, although it is located far away from the centroid. Whereas the members c_l belong to a second group, they will be transferred to other clusters, although it is located close to the centroid. This inappropriate transferring of members will also increase the clusters' intra-cluster variance and produce a poor segmentation.
- Furthermore, the MKM clustering has a pair of fitness conditions to avoid making an empty cluster in the clustering process (Siddiqui & Isa, 2011). However, the fitness conditions are not robust enough to distinguish between the empty clusters and the cluster with similar value members. This is because of similar fitness values (i.e., zero) for both types of clusters. Thus, transferring members to the cluster with similar value members causes the empty cluster to avoid updating the clustering process.

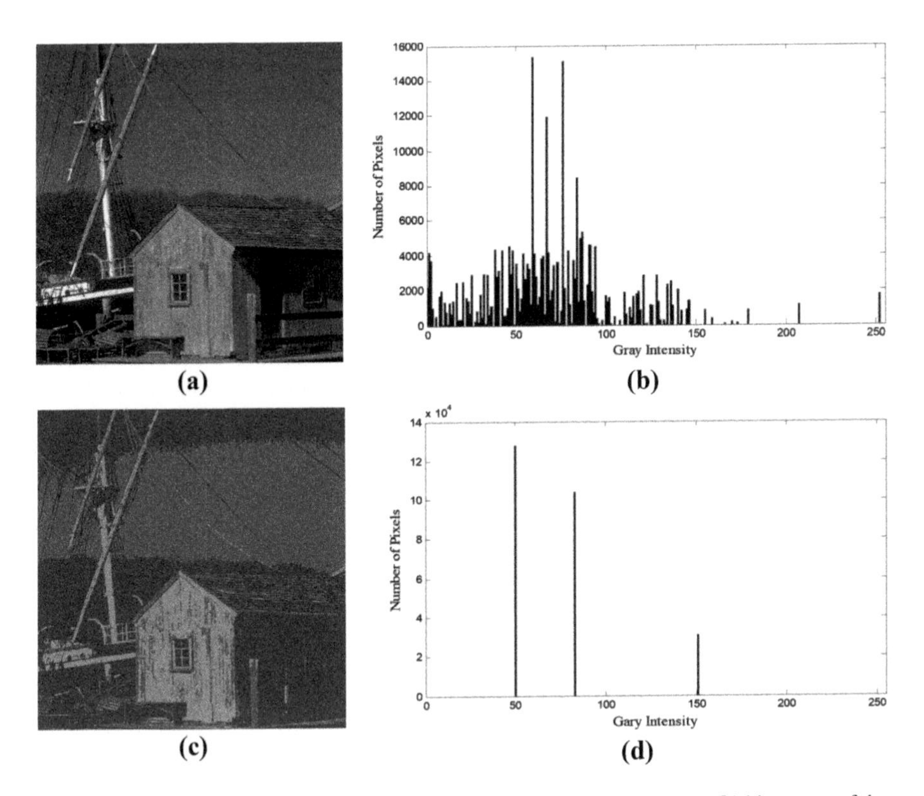

Fig. 2.5 Applying MKM Clustering on *Harbor House* image; (**a**) test image, (**b**) histogram of the test image, (**c**) segmented image, and (**d**) histogram of the segmented image

2.3.3 Illustration of MKM Clustering's Limitations

To illustrate the MKM clustering limitations, the *Harbor House* image is selected as a tested image (Fig. 2.5). The initial parameters settings of the MKM clustering are as follows: the number of clusters is set as three, and the constant value for $\alpha_a = \alpha_b = \alpha_o$ is set as 0.3. According to theory, each constant is set in the range of $0 < \alpha_a < 1/3$ (Mashor, 2000). The segmented image and its histogram after applying the MKM clustering are also depicted in Fig. 2.5 (c) and (d), respectively. Like k-means clustering, the MKM clustering also has been unsuccessful in segmenting the regions of interest with sharp edges and detailed information, as shown in Fig. 2.5 (c). For instance, the fence at the front of the house is not segmented. It is also confirmed from Fig. 2.5 (d), where the histogram presents the segmented image's inappropriate results. The centroids are not located at the prominent peaks

Table 2.2 Quantitative analysis of the MKM clustering using *Harbor House* image

Initialized number of clusters	Final number of clusters	MSE	INTER	*F(I)* 1x10^7	*F'(I)*	*Q(I)*
3	3	423.00	67.33	17.1935	13.286	230.667

of the histogram of the test image. For example, the centroid with an intensity value of 150 that located far from active region. To confirm this limitation, the quantitative analysis is performed, and the large value of the MSE (it represent intra-cluster variance) and the small value of the INTER (it represent inter-cluster variance) show the poor performance of MKM clustering (Table 2.2). Furthermore, the large value of the *F(I)*, *F′ (I)*, and *Q(I)* confirms the non-homogenous segmentation by the MKM clustering.

2.4 Adaptive Moving k-means Clustering

AMKM clustering has followed the MKM clustering principle with some changes in its pixels transferring approach (Isa et al., 2009). AMKM considers that transferring member pixels of the cluster with the highest fitness to the cluster with the lowest fitness value is the main reason for the MKM clustering to converge to the poor local minima. Therefore, AMKM clustering used the modified transferring process. The cluster's member pixels with the highest fitness value are transferred to the nearest cluster instead of the cluster with the lowest fitness value.

2.4.1 *Implementation of AMKM Clustering*

As compared to the MKM clustering, the AMKM clustering is more robust in assigning the member pixels to their respective cluster. It has not followed the conventional pixels transferring process, where certain members of the cluster with the largest fitness are forced to become cluster members with the smallest fitness. In the AMKM clustering, the cluster members with the largest fitness will be assigned to the nearest cluster instead of to the cluster with the smallest fitness. It is believed that a decrease in the inappropriate transferring process of member pixels can avoid the major problem of MKM clustering, i.e., segmentation with large intra-cluster variance. The AMKM clustering implementation is very much similar to the MKM clustering except for the transferring process. A pseudo-code of AMKM is as follow:

```
                where, n_{c_s} and n_{c_l} are the number of new members of c_s and c_l,
                respectively.
                // Update α_a according to α_a = α_a − α_a/k
        end if
    else
        // measure the distance between all p_i(x, y) of the images and the updated
        centroids.
        // p_i(x, y) with minimum distance assigned to their centroids. In other words,
        labeling the segmented image pixels with the centroid value of their nearest
        clusters.
        // measure the (n_j) number pixels in each cluster.
        // calculate the new value of centroids using Eq. 2.1.
        // measure the fitness value for all clusters using Eq. 2.2.
        // Update α_a and α_b according to α_a = α_o and α_b = α_b − α_b/k , respectively.
        if f(c_s) ≤ α_b f(c_l)
                // return to the fitness verification step of this clustering process
        else
                //jump to terminate all process, and return the segmented image and
                information centroid (c_j)
        end if
    end if
end while
```

The flow diagram for the implementation of the AMKM clustering is as shown in Fig. 2.6.

2.4.2 *Limitations of AMKM Clustering*

In the AMKM clustering, a transferring concept is evolved by that of the cluster's member pixels with the largest fitness value (c_l). Those that satisfy the defined condition $p_i(x, y) < c_l$ are moved to the nearest cluster instead of moving in the cluster with the smallest fitness value (c_s). Unlike the MKM clustering, the AMKM clustering does not allow the cluster c_s to withdraw all its members and becomes an empty cluster. This new approach makes AMKM efficient to avoid the inappropriate transferring of data between the clusters. However, the member transferring process's advanced approach encourages the trapped centroid problem. The cluster with a smaller fitness value, mostly located far away from other clusters, may not be updated in the next iteration of the clustering process (Siddiqui & Isa, 2011). Two more limitations of the MKM clustering, i.e., (a) inappropriate range selection of the c_l members for transferring between clusters and (b) the fitness condition for transferring process makes MKM clustering incompetent to differentiate the empty clusters and clusters with similar value member pixels (a.k.a. zero variance clusters).

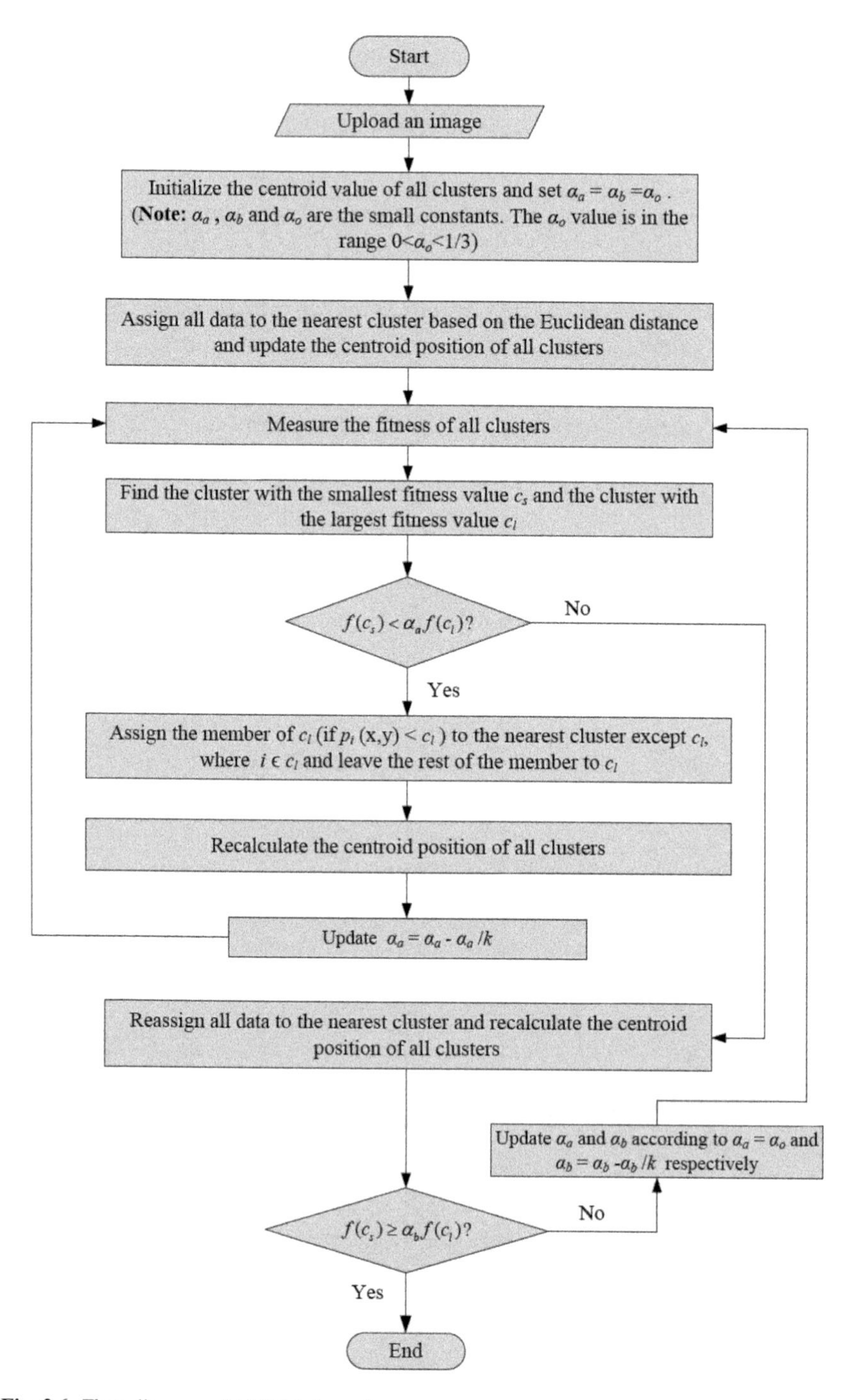

Fig. 2.6 Flow diagram of AMKM clustering

2.4.3 Illustration of AMKM Clustering's Limitations

The *Gantry Crane* image is selected as a test image to demonstrate the AMKM clustering limitations. Here, except for the number of clusters, the rest of the initial parameter settings are the same as defined earlier for the MKM clustering, and the value of the number of clusters is set to four. The test and segmented images with their histograms are shown in Fig. 2.7, where the centroid in a histogram of the segmented image, represented with its value 140, is trapped at a region far from one of the most active region (i.e., the region with the range 0 to 50 on intensity). Because of this trapped centroid, the AMKM clustering is failed to segment the switch panel on a rig structure, as shown in Fig. 2.7 (c). The quantitative analysis is

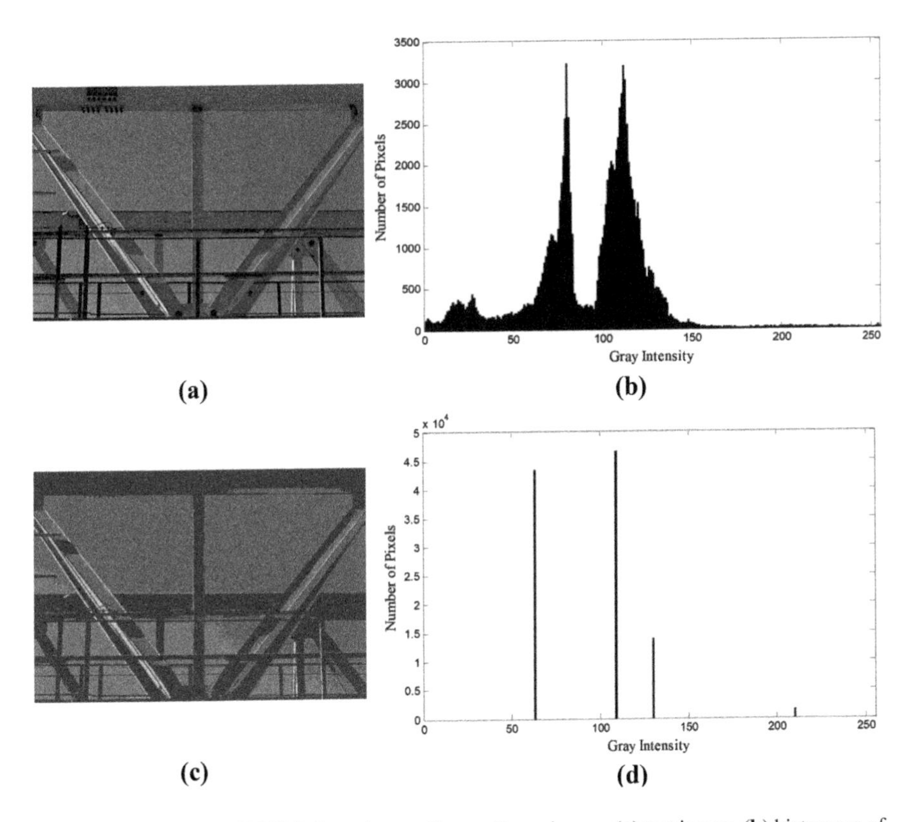

Fig. 2.7 Applying the AMKM clustering on *Gantry Crane* image; (**a**) test image, (**b**) histogram of the test image, (**c**) segmented image, and (**d**) histogram of the segmented image

Table 2.3 Quantitative analysis of the AMKM clustering using *Gantry Crane* image

Initialized number of clusters	Final number of clusters	MSE	INTER	$F(I)$ 1×10^7	$F'(I)$	$Q(I)$
4	4	258.194	77	2.707	3.558	36.971

tabulated in Table 2.3 to illustrate the AMKM clustering performance on the test image. The large MSE value and small INTER value confirm the centroids' false location, i.e., the non-active regions. Whereas the large value of the $F(I)$, $F'(I)$, and $Q(I)$ function shows the poor segmentation (i.e., image is segmented in non-homogenous regions) result of the AMKM clustering. All this occurs because of the trapped centroid in a non-active region.

2.5 Fuzzy c-means Clustering

In contrast to the k-means clustering, the fuzzy c-means clustering has soft membership by which each pixel partially belongs to each cluster instead of completely becoming a part of a single cluster. Its main aim is to iteratively minimize the defined objective function (Bezdek, 1981). The fuzzy c-means clustering starts with the assumption of each pixel's membership value related to all clusters. Next, each cluster's centroid is measured by taking an average of all the pixels with their different degree (it is determined in each iteration by executing the soft membership function). This is an iterative process and continues until no change is observed in all pixels' membership values related to the clusters. The fuzzy concept makes the fuzzy c-means clustering more flexible to keep the centroids as close as possible to active regions and become less sensitive to initial settings. However, the fuzzy c-means clustering has no specific boundary between the clusters to avoid clusters' overlapping (Sulaiman & Isa, 2010).

2.5.1 Implementation of Fuzzy c-means Clustering

The fuzzy c-means clustering segments the pixels of an image into overlapping regions. The fuzzy c-means clustering can be implemented as follow:

```
input: test image with the size of N × M
input: k number of clusters
input: generate a random value of the centroids (c_i)
input: set α_a = α_b = α_o. (here, α_a, α_b and α_o are the small constants. The α_o value is in the
range 0 < α_o < 1/3).
output: centroids value and segmented image

initialize
// generate a segmented image with the test image's size, and it contains no information
on their pixels.
while
    for i ← 1 to N do
        for j ← 1 to M do
            // measure the distance between all p_i(x, y) of the test images and
            centroids.
            // p_i(x, y) with minimum distance assigned to their centroids. In other
            words, labeling the segmented image pixels with the centroid value
            of their nearest clusters.
            // measure the (n_j) number pixels in each cluster.
        end for
    end for
    // calculate the new value of centroids using Eq. 2.1.
    // measure the fitness value for all clusters using Eq. 2.2.
    // find the cluster with the smallest and largest fitness value, i.e., c_s and c_l,
    respectively.
    if f(c_s) < α_a f(c_l)                    */ fitness verification step
        // analyzed p_i(x, y) of segmented image association with clusters
        if p_i(x, y) < c_l, here i ∈ c_l
            // Assign those pixels c_l to the nearest cluster and leave the rest of the
            pixels in c_l.
            // Recalculate the centroid position of all clusters using Eq. 1.
            // Update α_a according to α_a = α_a − α_a/k
        end if
    else
        // measure the distance between all p_i(x, y) of the test images and the
        updated centroids.
        // p_i(x, y) with minimum distance assigned to their centroids. In other words,
        labeling the segmented image pixels with the centroid value of their nearest
        clusters.
        // measure the (n_j) number pixels in each cluster.
        // calculate the new value of centroids using Eq. 2.1.
        // measure the fitness value for all clusters using Eq. 2.2.
        if f(c_s) ≥ α_b f(c_l)
```

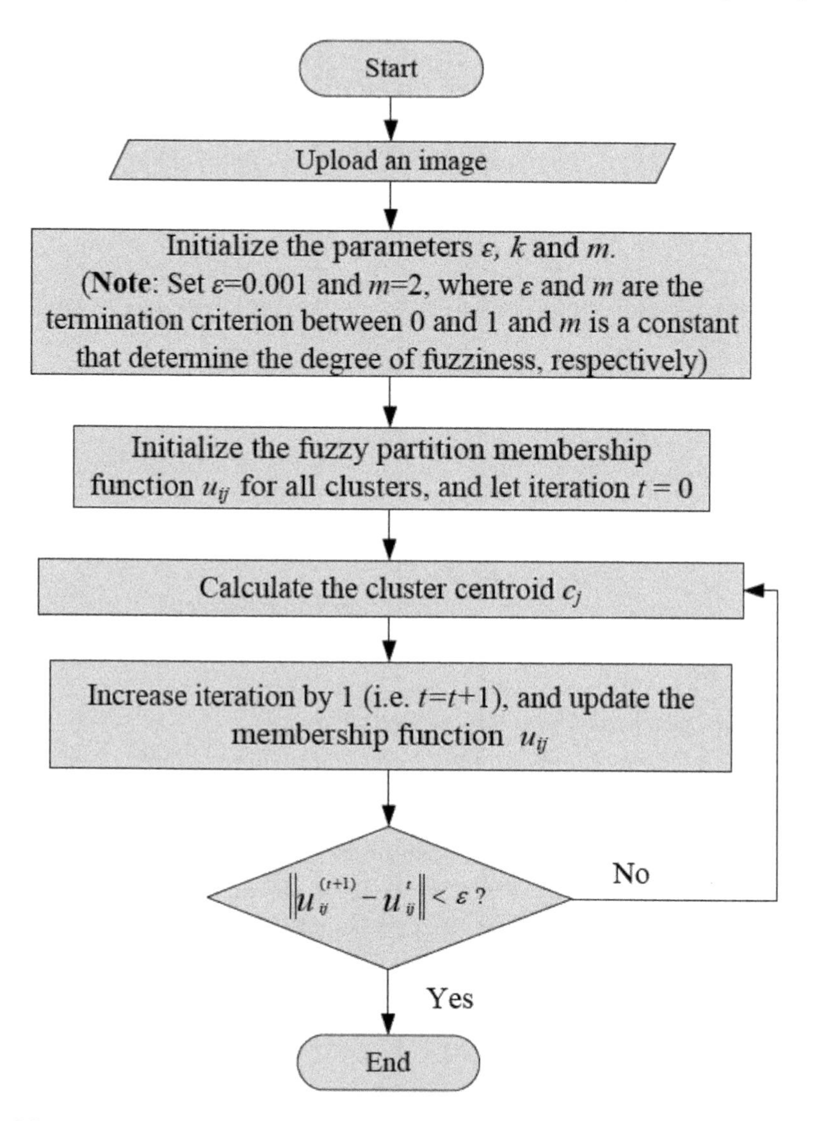

Fig. 2.8 Flow diagram of fuzzy c-means clustering

The flow chart for the implementation of the fuzzy c-means clustering is shown in Fig. 2.8.

2.5.2 *Limitations of Fuzzy c-means Clustering*

In the fuzzy c-means clustering, each pixel has partially belonged to all clusters with some degree, i.e., a pixel's membership value for clusters. The membership is measured by using equation 2.6. The fuzzy characteristic of fuzzy c-means restricts the membership value of a pixel is in the range between 0 and 1, and the sum of membership value of each pixel in all clusters is equal to 1 (Siddiqui et al., 2013). This characteristic gives a sufficient membership value for the outliers (far located pixels from the centroid) to become a cluster member, increases the intra-cluster variance, and reduces the inter-cluster variance (Thomas et al., 2009). Besides, the increase of overlapping between the clusters in the data causes the increase of sensitivity of fuzzy c-means to outliers (Dixon et al., 2009; Yang et al., 2004). (Kersten, 1999) Euclidean distance for measuring the membership value in the fuzzy c-means clustering is sensitive to outliers. It is confirmed in another research (Hathaway et al., 2002). It is considered that the Euclidean distance assigned a high membership to outliers, which indirectly pulls the cluster from the optimum location.

Besides, the convergence speed of fuzzy c-means clustering is too slow in comparison with the k-means clustering. An earlier study on fuzzy clustering (Wei & Xie, 2000) presents the convergence speed comparison between the fuzzy c-means clustering and the introduced clustering called Revival Check Fuzzy C-Means (RCFCM) clustering. The RCFCM clustering speeds up the fuzzy clustering process's convergence speed by magnifying the biggest membership of a pixel and suppressing the second biggest membership of a pixel. But it is found in the literature that this approach often is not converged at the optimum global location (Fan et al., 2003). The main problem is ignoring the pixel memberships, not the two leading highest pixels' highest memberships. Besides, a parameter α has a significant role in controlling the clustering process. Any improper selection of α value can disturb the modification of the second-highest membership and ordering of membership. The unbalanced ordering of memberships leads the final solution to converges at poor local minima. Another modification in fuzzy clustering, called the Suppress-Fuzzy C-means (S-FCM) clustering, was proposed by Fan et al. (2003). It magnified the biggest membership without disturbing the order of memberships. S-FCM clustering has also employed a controlling parameter α. Its value equal to 0 converts the SFCM clustering into the k-means clustering; its value equal to 1 converts the S-FCM into fuzzy clustering (Fan et al., 2003). The performance of the S-FCM clustering only relies on the personal setting of α value.

2.5.3 *Illustration of Fuzzy c-means Clustering's Limitations*

In fuzzy c-means clustering, the pixels become members of the cluster with the highest membership value; the ideal condition is that the membership value is equal to 1 or greater than the sum of membership value of pixels with other clusters (Siddiqui et al., 2013). However, in some cases, the pixels located between two clusters have a reasonable degree of membership value with faraway clusters; these pixels act as outliers for those clusters. This phenomenon has typically occurred if a range of data is distributed between more than two neighboring clusters. The fuzzy c-means clustering applied to segment the manually generated data of intensity ranged between 1 and 120 into three regions (Siddiqui et al., 2013). In the segmented result of fuzzy c-means clustering, the range of data divides into three regions (i.e., *C1*, *C2*, and *C3*) with their membership functions (i.e., *u1*, *u2*, and *u3*, respectively). From Fig. 2.9, the membership functions are represented with exclusive line textures and colors, i.e., green dotted (*u1*), red dash (*u2*), and blue solid (*u3*) lines. For instance, the membership function *u1,* where the data ranged between 0 to 60, is located far from cluster *C1* but still received the reasonable membership value to become outliers for cluster *C1*. A similar pattern is observed in other membership functions, i.e., *u2* and *u3*. Therefore, these outliers can pull away from the centroids from their active regions and generate clusters with large intra-cluster variance and low inter-cluster variance.

In Fig. 2.10, a *Football* image is selected as a test image to illustrate outliers' effect on fuzzy c-mean clustering. In initial settings, the number of clusters and termination criterion is set to 3 and 0.001, respectively. Histogram information in Fig. 2.10 (b) recommends that three final centroids' ideal locations in the intensity scale are 50, 140, and 255. But the two final centroids are located very close to each other (i.e., 50 and 77), and the region with intensity value closed to 255 (i.e., pixels of football laces) is not segmented. This poor representation of data is because of the outliers that localize the centroids at false or non-active regions. The quantitative analysis also confirms the poor segmentation or poor convergence of the final solution. The large value of the MSE and VXB and the small value of the INTER are obtained. As a tabulated result in Table 2.4, the large value of *F(I)*, *F' (I)*, and *Q(I)* functions confirms non-homogeneous segmentation of the image by fuzzy c-means clustering.

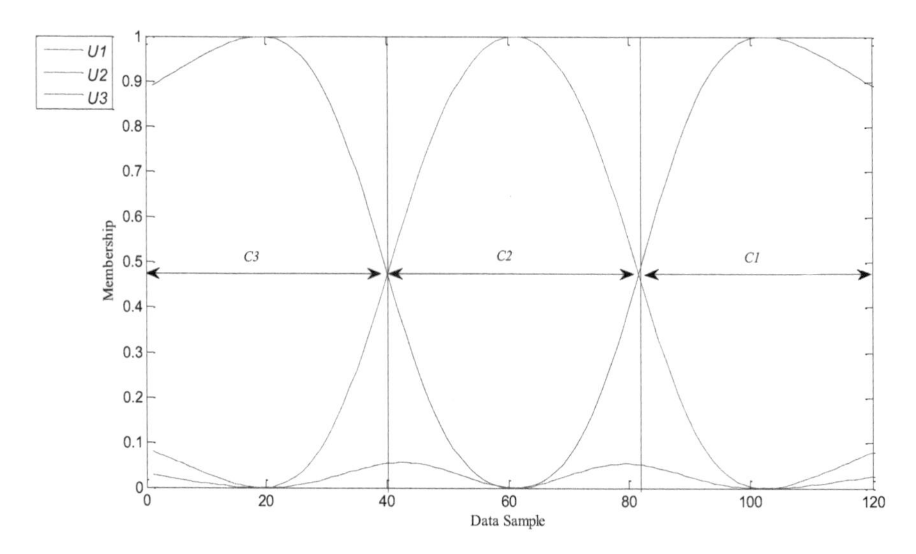

Fig. 2.9 Mapping of membership functions of fuzzy c-means clustering using data ranged between 1 and 120. (Siddiqui et al., 2013).

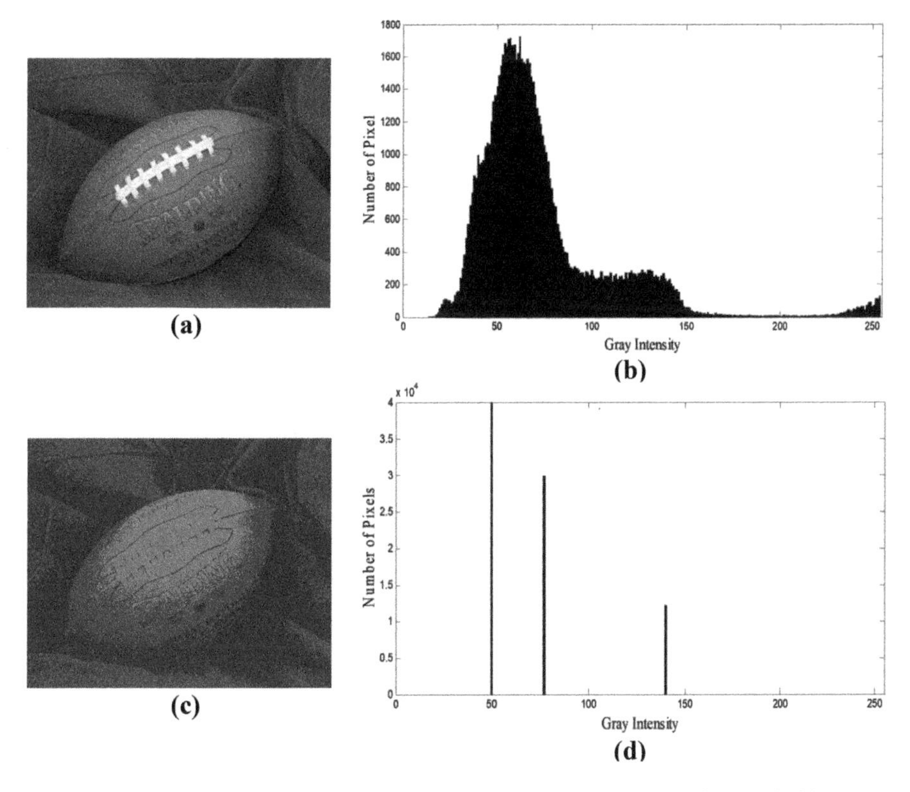

Fig. 2.10 Applying the fuzzy c-means clustering on *Football* image; (**a**) test image, (**b**) histogram of the test image, (**c**) segmented image, and (**d**) histogram of the segmented image

Table 2.4 Quantitative analysis of fuzzy c-means clustering

Initialized number of clusters	Final number of clusters	MSE	INTER	VXB	$F(I)$ $1\text{x}10^7$	$F'(I)$	$Q(I)$
3	3	339.69	60	0.5132	4.238	13.686	917.22

2.6 Adaptive Fuzzy Moving k-means Clustering

Adaptive fuzzy moving k-means (AFMKM) clustering was introduced for image segmentation (Isa et al., 2010). In AFMKM clustering, the cluster members' transferring concept of the AMKM and fuzzy connect of the fuzzy c-means clustering are integrated to improve the transferring processes. The pixels with mass (i.e., the fuzzy weight of a pixel in a cluster) less than the centroid value of the cluster with the highest fitness are moved from the cluster with the highest fitness value to the nearest cluster (Isa et al., 2010). Based on the study (Sulaiman & Isa, 2010), this approach significantly produces better segmentation performance than AMKM clustering.

2.6.1 Implementation of Adaptive Fuzzy Moving k-means Clustering

The AFMKM clustering is the modified version of the AMKM clustering, which incorporates the fuzzy concept. For illustration, the complete implementation of the AFMKM clustering is as follow:

```
                    // Update α_a and α_b according to α_a = α_o and α_b = α_b − α_b/k ,
                    respectively.
                    // return to the fitness verification step of this clustering process
            else
                    //jump to terminate all process, and return the segmented image and
                    information centroid (c_j)
            end if
        end if
end while
```

The flow diagram of the AFMKM clustering is shown in Fig. 2.11.

2.6.2 Limitations of Adaptive Fuzzy Moving k-means Clustering

In the AFMKM clustering, the fuzzy concept of fuzzy c-means clustering is fused with the transferring concept of MKM clustering. As it is a well-proven argument, the fuzzy concept is mainly sensitive to outliers (Hathaway et al., 2002) and leads the solution to converge at the non-active region or poor local minima; the fuzzy concept makes the AFMKM clustering highly sensitive to outliers as well. AFMKM has two more limitations because of incorporating the MKM clustering: (a) transferring of unsuitable range of members from the cluster with largest fitness value toward nearest clusters and (b) fitness condition itself is not capable of distinguishing the clusters with members of similar value and empty clusters (Siddiqui & Isa, 2011).

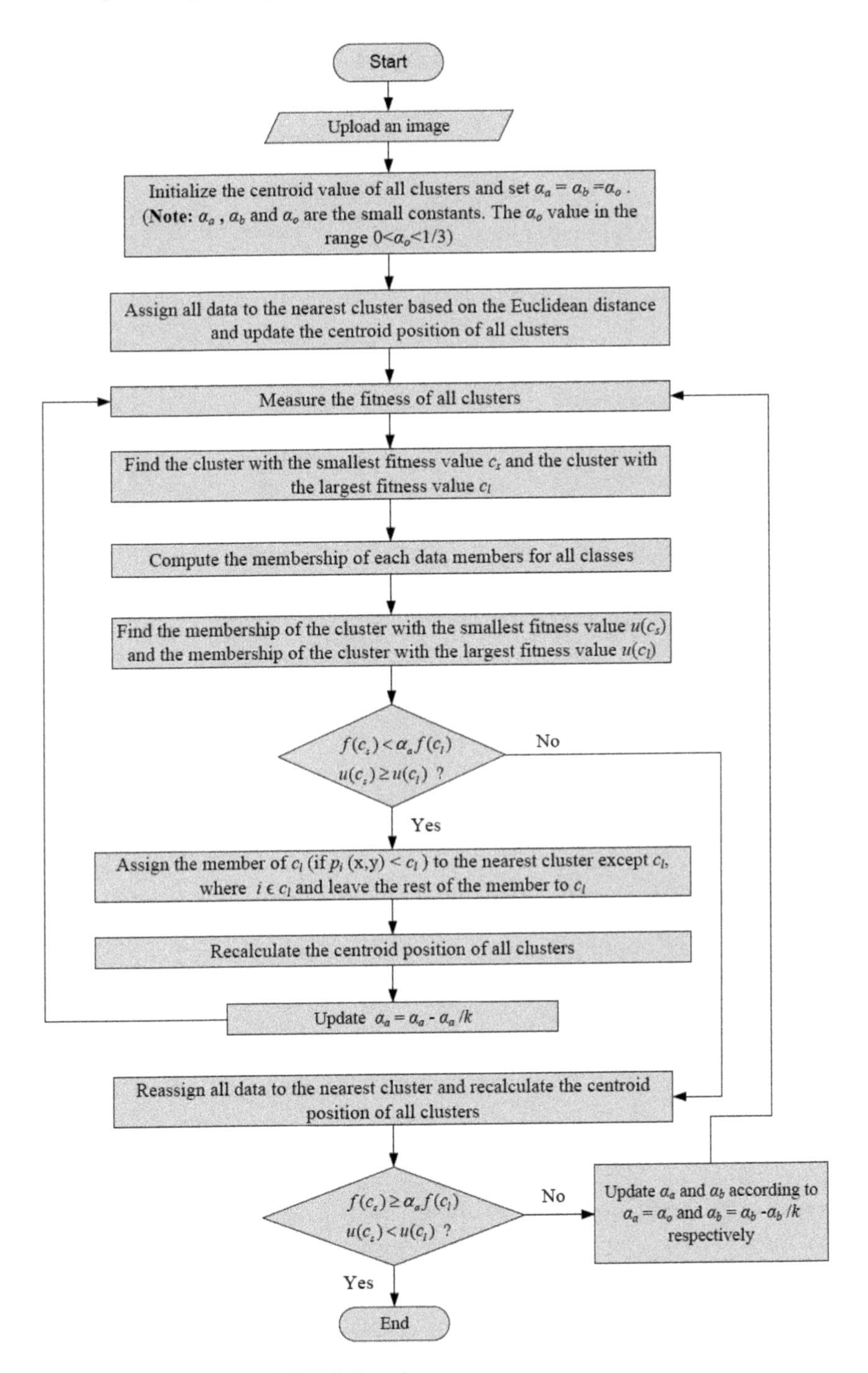

Fig. 2.11 Flow diagram of AFMKM clustering

2.6.3 *Illustration of Adaptive Fuzzy Moving k-means Clustering's Limitations*

For illustrating the limitations of AFMKM clustering, the *Light House* image is selected as the test image. In initial settings, the number of clusters is set to three. Histogram information in Fig. 2.12 (b) suggests three centroids' ideal location in the intensity scale is 80, 130, and 250. Based on Fig. 2.12 (d), the AFMKM

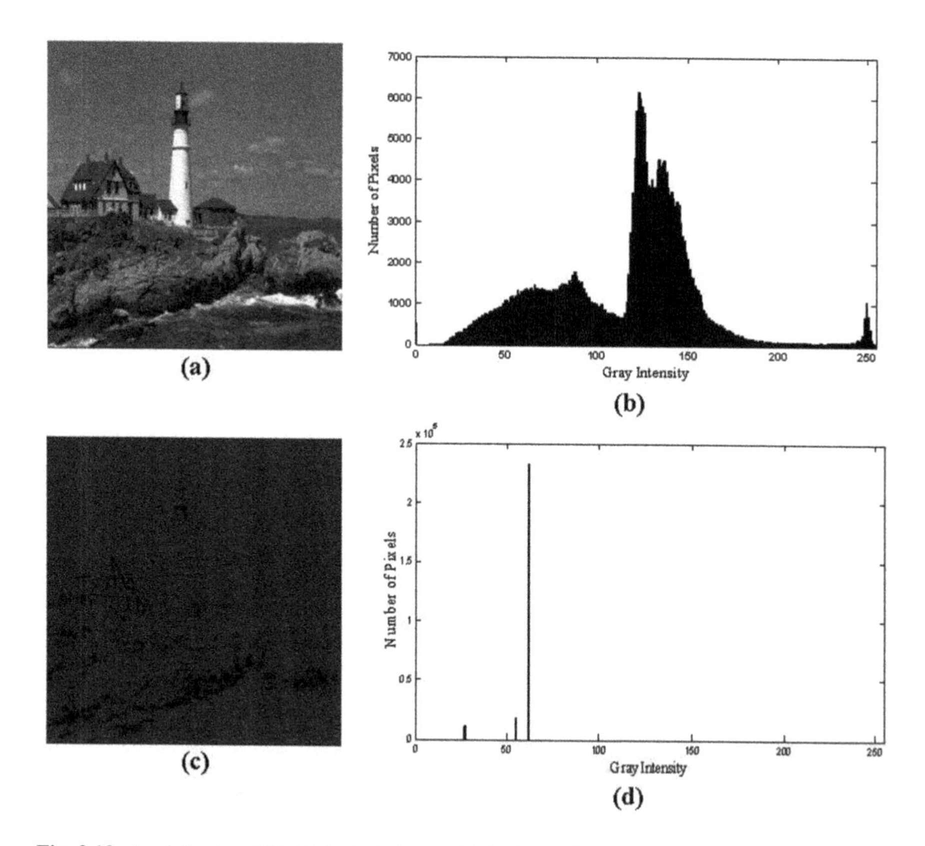

Fig. 2.12 Applying the AFMKM clustering on *Light House* image; (**a**) test image (**b**) histogram of the test image, (**c**) segmented image, and (**d**) histogram of the segmented image

Table 2.5 Quantitative analysis of the AFMKM clustering using *Light House* image

Initialized number of clusters	Final number of clusters	MSE	INTER	VXB	*F(I)* 1×10^7	*F'(I)*	*Q(I)*
3	3	4599.9	23.33	9.154	15.28	12.99	382.8

generates centroids located at 25, 55, and 60. Because of false centroids localization, the incomplete shape of regions with blur edges are segmented (Fig. 2.12 (c)). Based on a tabulated quantitative analysis in Table 2.5, the large value of the MSE, *F(I)*, *F' (I)*, and *Q(I)* and small value of INTER confirm the empty cluster and trapped centroid limitations of AFMKM clustering.

2.7 Adaptive Fuzzy k-means Clustering

Sulaiman and Isa (2010) proposed the adaptive fuzzy K-means (AFKM) clustering as an advanced version of AFKM clustering that fused a fuzzy concept with a novel belongingness concept. Sulaiman and Mat Isa (2010) considered that membership alone is not the best approach for assigning pixels to clusters. Therefore, a novel belongingness concept is derived from the membership function of fuzzy clustering to ensure a strong relationship between the clusters and their members.

2.7.1 Implementation of Adaptive Fuzzy K-means Clustering

As discussed earlier, the AFKM clustering integrates the belongingness concept and fuzzy concept to strengthen the relationship between clusters and their members. Based on the literature (Sulaiman & Isa, 2010), AFKM clustering is implemented as follow:

input: test image with the size of $N \times M$
input: define the value of termination criteria (set within the range of 0 and 1 and here $\varepsilon = 0.001$) and degree of fuzziness (e.g., $m = 2$)
input: k number of clusters
input: randomly generate membership function u_{ij} for all clusters.
output: centroids value and segmented image

initialize
// generate a segmented image with the test image's size, and it contains no information on their pixels.
while
  for $i \leftarrow 1$ **to** N **do**
    for $j \leftarrow 1$ **to** M **do**
      // calculate the cluster centroid c_j according to Eq. 2.5.

$$c_j = \frac{\sum_{i=1}^{n} \left(u_{ij}^m \times p_i(x,y) \right)}{\sum_{i=1}^{n} u_{ij}^m} \quad \textbf{(2.5)}$$

      // $p_i(x,y)$ with the highest membership are assigned to their respective centroids. In other words, the segmented image pixels are labeled with the centroid value of their clusters.
      // measure the distance between all $p_i(x,y)$ of the test images and centroids.
      //compute the new membership function u_{ij} based on Eq. 2.6.

$$u_{ij} = \frac{1}{\sum_{h=1}^{k} \left[\frac{\left\| p_i(x,y) - c_j \right\|}{\left\| p_i(x,y) - c_h \right\|} \right]^{2/(m-1)}} \quad \textbf{(2.6)}$$

    end for
  end for
  if $\left\| u_{ij}^{(t+1)} - u_{ij}^{t} \right\| < \varepsilon$
    // terminate the process.
  end if
  // erase all information from the segmented image and c_j.
end while

input: test image with the size of $N \times M$
input: define the value of m (degree of fuzziness)
input: k number of clusters
input: randomly generate membership function u_{ij} for all clusters.
input: generate a random value of the centroids (c_i)
input: set $\alpha_a = \alpha_b = \alpha_o$. (here, α_a, α_b and α_o are the small constants. The α_o value is in the range $0 < \alpha_o < 1/3$).
output: centroids value and segmented image

initialize
// generate a segmented image with the test image's size, and this image contains no information on their pixels.
while
  for $i \leftarrow 1$ **to** N **do**
    for $j \leftarrow 1$ **to** M **do**
      // measure the distance between all $p_i(x, y)$ of the test images and centroids.
      // $p_i(x, y)$ with minimum distance assigned to their centroids. In other words, labeling the segmented image pixels with the centroid value of their nearest clusters.
      // measure the (n_j) number pixels in each cluster.
    end for
  end for
  // calculate the new value of centroids using Eq. 2.1.
  // measure the fitness value for all clusters using Eq. 2.2.
  // find the cluster with the smallest and largest fitness value, i.e., c_s and c_l, respectively.
  //compute the new membership function u_{ij} based on Eq. 2.6.
  // find the cluster's membership with the smallest and largest fitness value, i.e., $u(c_s)$ and $u(c_s)$, respectively.
  if $f(c_s) < \alpha_a f(c_l)$ & $u(c_s) \geq u(c_l)$ ***/ fitness verification step**
    // analyzed $p_i(x, y)$ of segmented image association with clusters
    if $p_i(x, y) < c_l$, here $i \in c_l$
      // Assign those pixels c_l to the nearest cluster and leave the rest of the pixels in c_l.
      // Recalculate the centroid position of all clusters using Eq. 2.1.
      // Update α_a according to $\alpha_a = \alpha_a - \alpha_a / k$
    end if
    // measure the distance between all $p_i(x, y)$ of the test images and the updated centroids.

```
            // p_i(x, y) with minimum distance assigned to their centroids. In other words,
            labeling the segmented image pixels with the centroid value of their nearest
            clusters.
            // measure the (n_j) number pixels in each cluster.
            // calculate the new value of centroids using Eq. 2.1.
            // measure the fitness value for all clusters using Eq. 2.2.
            if f(c_s) ≥ α_b f(c_l) & u(c_s) < u(c_l)
                  // Update α_a and α_b according to α_a = α_o and α_b = α_b − α_b/k ,
                  respectively.
                  // return to the fitness verification step of this clustering process
            else
                  //jump to terminate all process and return the segmented image and
                  information of centroid (c_j)
            end if
        end if
end while
```

input: test image with the size of $N \times M$
input: define the value of m (degree of fuzziness)
input: k number of clusters
input: generate a random value of the centroids (c_i)
input: randomly generate membership function u_{ij} for all clusters.
output: centroids value and segmented image

initialize
// generate a segmented image with the test image's size, and this image contains no information on their pixels.
while
for $i \leftarrow 1$ **to** N **do**
for $j \leftarrow 1$ **to** M **do**
// calculate the cluster centroid c_j according to Eq. 2.5.
// $p_i(x, y)$ with the highest membership are assigned to their respective centroids. In other words, the segmented image pixels are labeled with the centroid value of their clusters.
// measure the distance between all $p_i(x, y)$ of the test images and centroids.
//compute the new membership function u_{ij} based on Eq. 2.6.
end for
end for
// calculate the degree of belongingness B_j of pixels to all clusters using equation

$$B_j = \frac{c_j}{u_{ij}^m} \quad \textbf{(2.7)}$$

// measure the e_{ij} using equation 8, where, B'_j is the normalized value of B_j

$$e_j = B_j - B'_j \quad \textbf{(2.8)}$$

// update membership using equation 2.9.

$$(u_{ij}^m)' = u_{ij} + (\Delta u_{ij}^m)' \quad \textbf{(2.9)}$$

where $(\Delta u_{ij}^m)' = \alpha (c_j)(e_j)$ **(2.10)**

here, α is a constant with a value between 0 and 1 and typically set to 0.1
// update centroid based on equation 2.11.

$$c_j = \frac{\sum_{i=1}^{n} \left((u_{ij}^m)' \times p_i(x, y) \right)}{\sum_{i=1}^{n} (u_{ij}^m)'} \quad \textbf{(2.11)}$$

if $c_j \neq c_i$
// $c_i = c_j$
else
// terminate the clustering process
end if
end while

The flow diagram of the AFKM clustering is shown in Fig. 2.13.

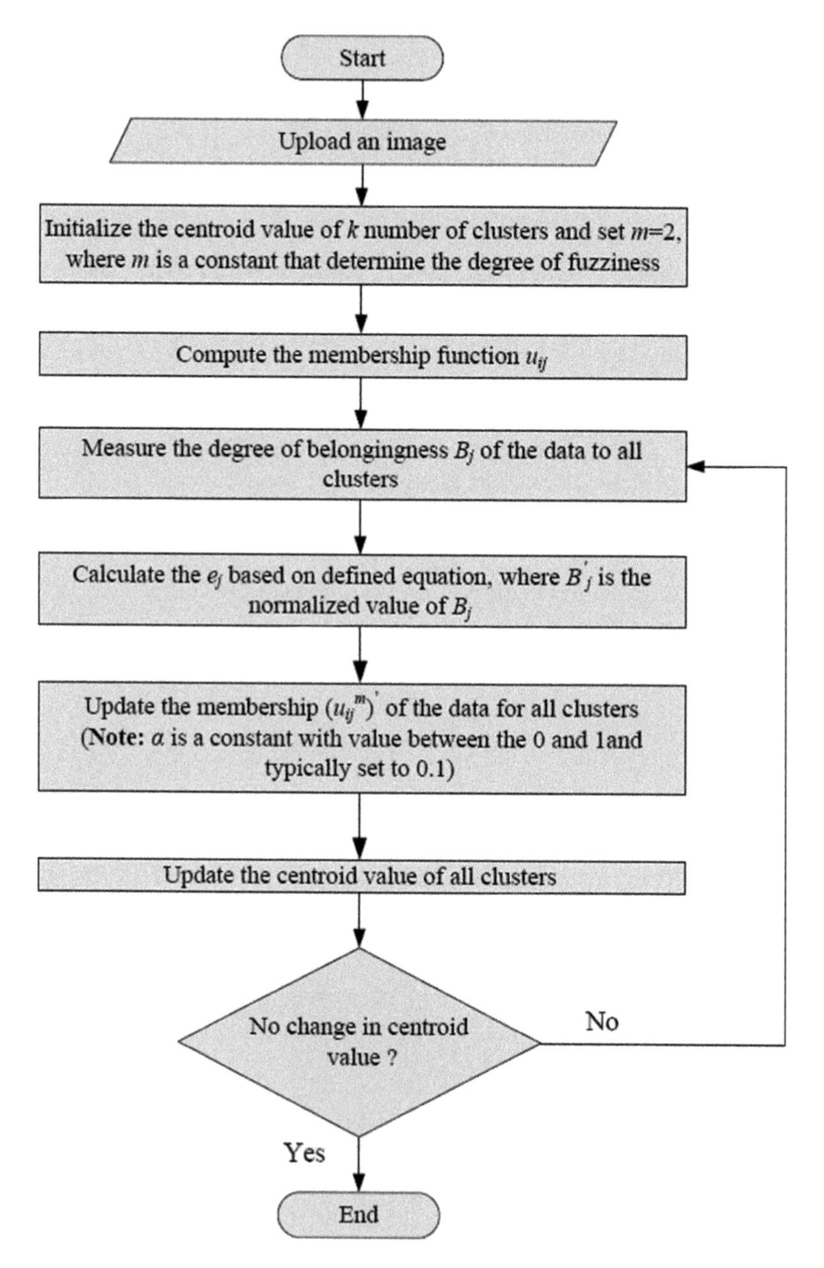

Fig. 2.13 Flow diagram of AFKM clustering

2.7.2 *Limitations of Adaptive Fuzzy k-means Clustering*

The belongingness concept of AFKM clustering is derived from the membership (i.e., fuzzy concept) and is simply the opposite of a fuzzy membership evaluation. Therefore, the AFKM clustering confronts limitations such as the empty cluster and centroid trapping at poor local minima. Besides, AFKM is also highly sensitive to outliers (Hathaway et al., 2002), which leads the AFKM to converge at the poor local minima.

2.7.3 *Illustration of Adaptive Fuzzy k-means Clustering's Limitations*

The *Monument Building* image is selected as a test image to explain AFKM clustering's limitation. In the initial setting, the number of clusters is set to four. However, the image is segmented into three clusters. It is also confirmed from the tabulated results in Table 2.6. The histograms of the test image and the segmented image are shown in Fig. 2.14 (b) and (d), respectively. It confirms the formation of an empty cluster in the segmented result as shown in Fig. 2.14 (d), where only three centroids instead of four centroids are obtained. Also, the segmented image is completely blurred; much useful information is failed to segment by the AFKM clustering (Fig. 2.14 (c)). The quantitative results of AFKM clustering are tabulated in Table 2.6, where the large values of MSE, *F(I)*, *F' (I)*, and *Q(I)* confirm the poor convergence property of the AFKM clustering. The infinity value of VXB proves that the clustering process end with two centroids having similar intensity. One cluster with similar intensity holds most of the pixels, while the other cluster with similar intensity has no pixels and becomes an empty cluster.

Table 2.6 Quantitative analysis of the AFKM clustering using '*Monument Building* image

Initialized number of clusters	Final number of clusters	MSE	INTER	VXB	*F(I)* 1×10^7	*F'(I)*	*Q(I)*
4	3	1217.6	52.16	∞	31.22	22.66	330.9

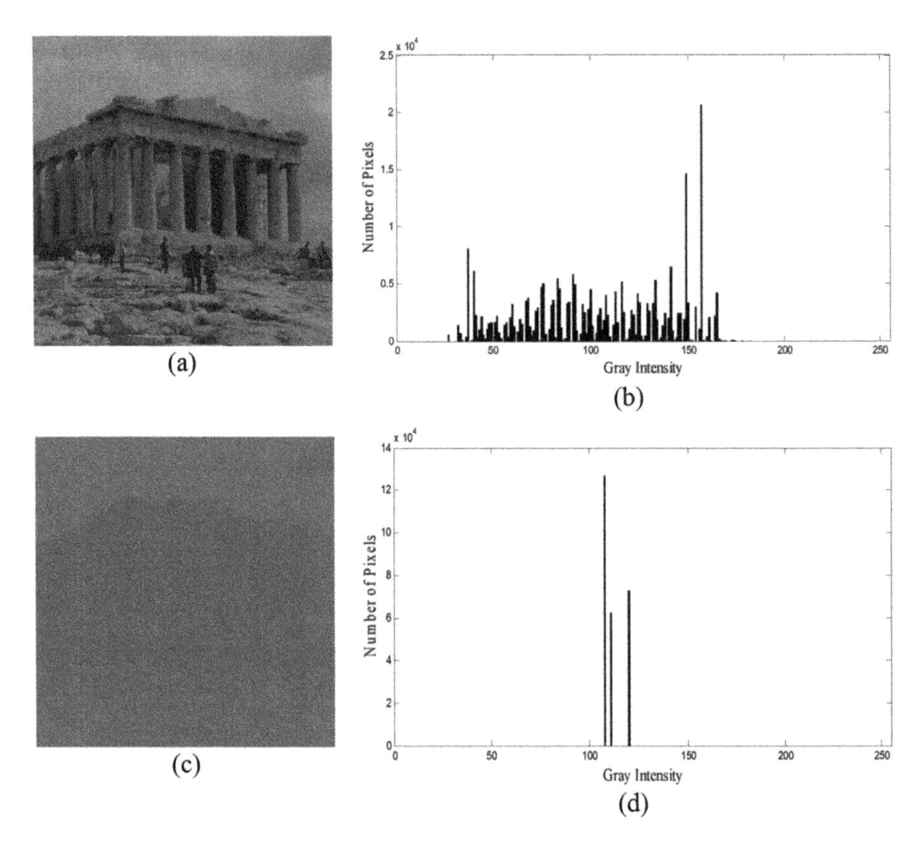

Fig. 2.14 Applying the AFKM clustering on the *Monument Building* image; (**a**) test image (**b**) histogram of the test image, (**c**) segmented image, and (**d**) histogram of the segmented image

References

Anderson, B. J., Gross, D. S., Musicant, D. R., Ritz, A. M., Smith, T. G., & Steinberg, L. E. (2006). *Adapting k-medians to generate normalized cluster centers.* Paper presented at the Proceedings of the 2006 SIAM International Conference on Data Mining.

Barioni, M. C. N., Razente, H. L., Traina, A. J. M., & Traina, C., Jr. (2008). Accelerating k-medoid-based algorithms through metric access methods. *Journal of Systems and Software, 81*(3), 343–355. https://doi.org/10.1016/j.jss.2007.06.019

Bezdek, J. (1981). *Pattern recognition with fuzzy objective function algorithms.* Plenum Press.

Bradley, P. S., & Fayyad, U. M. (1998). *Refining initial points for k-means clustering.* Paper presented at the ICML.

Dixon, S. J., Heinrich, N., Holmboe, M., Schaefer, M. L., Reed, R. R., Trevejo, J., & Brereton, R. G. (2009). Use of cluster separation indices and the influence of outliers: Application of two new separation indices, the modified silhouette index and the overlap coefficient to simulated data and mouse urine metabolomic profiles. *Journal of Chemometrics: A Journal of the Chemometrics Society, 23*(1), 19–31.

Ester, M., Kriegel, H.-P., & Xu, X. (1995). *A database interface for clustering in large spatial databases.* Inst. für Informatik.

Fan, J.-L., Zhen, W.-Z., & Xie, W.-X. (2003). Suppressed fuzzy c-means clustering algorithm. *Pattern recognition letters, 24*(9-10), 1607–1612.

Hathaway, R., Bezdek, J., & Hu, Y. (2002). Generalized fuzzy c-means clustering strategies using Lp norm distances. *IEEE Transactions on Fuzzy Systems, 8*(5), 576–582.

Isa, N. A. M., Salamah, S. A., & Ngah, U. K. (2009). Adaptive fuzzy moving K-means clustering algorithm for image segmentation. *IEEE Transactions on Consumer Electronics, 55*(4), 2145–2153.

Isa, N., Salamah, S., & Ngah, U. (2010). Adaptive fuzzy moving K-means clustering algorithm for image segmentation. *IEEE Transactions on Consumer Electronics, 55*(4), 2145–2153.

Kaufman, L., & Rousseeuw, P. J. (1990). Partitioning around medoids (program pam). *Finding groups in data: an introduction to cluster analysis, 344*, 68–125.

Kaufman, L., & Rousseeuw, P. J. (2009). *Finding groups in data: An introduction to cluster analysis* (Vol. 344). Wiley.

Kersten, P. R. (1999). Fuzzy order statistics and their application to fuzzy clustering. *IEEE Transactions on Fuzzy Systems, 7*(6), 708–712.

MacQueen, J. (1967). Some methods for classification and analysis of multivariate observations. In L. M. LeCam & J. Neyman (Eds.), *Proceedings of the fifth Berkeley symposium on mathematical statistics and probability*.

Mashor, M. (2000). Hybrid training algorithm for RBF network. *International Journal of The Computer, The Internet and Management, 8*(2), 50–65.

Siddiqui, F. U., & Isa, N. A. M. (2011). Enhanced moving K-means (EMKM) algorithm for image segmentation. *IEEE Transactions on Consumer Electronics, 57*(2), 833–841.

Siddiqui, F., & Isa, N. M. (2012). Optimized K-means (OKM) clustering algorithm for image segmentation. *Opto-Electronics Review, 20*(3), 216–225.

Siddiqui, F. U., Isa, N. A. M., & Yahya, A. (2013). Outlier rejection fuzzy c-means (ORFCM) algorithm for image segmentation. *Turkish Journal of Electrical Engineering & Computer Sciences, 21*(6).

Sulaiman, S. N., & Isa, N. A. M. (2010). Adaptive fuzzy-K-means clustering algorithm for image segmentation. *IEEE Transactions on Consumer Electronics, 56*(4), 2661–2668.

Thomas, B., Raju, G., & Sonam, W. (2009). A modified fuzzy c-means algorithm for natural data exploration. *World Academy of Science, Engineering and Technology, 49*.

Wei, L.-m., & Xie, W.-x. (2000). Rival checked fuzzy c-means algorithm. *Acta Electronica Sinica, 28*(7), 63–66.

Weisstein, E. W. (2004). K-Means Clustering Algorithm. *MathWorld–A Wolfram Web Resource.*

Yang, M.-S., Hwang, P.-Y., & Chen, D.-H. (2004). Fuzzy clustering algorithms for mixed feature variables. *Fuzzy Sets and Systems, 141*(2), 301–317.

Chapter 3
Novel Partitioning Clustering

3.1 Robust Techniques of Partitioning Clustering

Although the partitioning clustering techniques can simplify an image with less complexity, they have major problems that lead the final solution of clustering techniques to converge at poor local minima. These problems include initialization, trapped centroid at the non-active region, empty cluster, and outlier sensitivity. This chapter addresses the modifications in clustering techniques proposed in the literature by the authors of this book (Siddiqui & Isa, 2012; Siddiqui & Isa, 2011; Siddiqui et al., 2013) to overcome all the mentioned limitations.

The k-means and fuzzy c-means clustering techniques represent the hard-membership function-based clustering and fuzzy-membership function-based clustering techniques, respectively. The pixel assigning process of k-means was modified in the optimized k-means clustering (Siddiqui & Isa, 2012), and the working steps of this technique are discussed in detail in this chapter. It also covers the working steps of the enhanced moving k-means clustering technique (Siddiqui & Isa, 2011), where the transferring process of moving k-means clustering technique is modified to avoid empty cluster, dead centroid, and centroid trapping at non-active region problems of k-means clustering. Finally, the working steps of outlier rejection fuzzy c-means clustering technique are discussed in detail. This technique reduced the outlier sensitivity of fuzzy c-means clustering. Overall, the modifications made in the mentioned clustering techniques lead to the techniques' final solution at the optimum global location. Few samples image will be used in this chapter to discuss the performance of the modified clustering techniques.

F.U. Siddiqui, A. Yahya, *Clustering Techniques for Image Segmentation*,
https://doi.org/10.1007/978-3-030-81230-0_3

3.2 Optimized k-means Algorithm

Like k-means clustering, the OKM clustering assigns pixels to their nearest clusters using the Euclidean distance. But OKM has advanced the pixel process having the same Euclidean distance to two or more adjacent clusters. Such pixels are known as conflict pixels. The cluster fitness value is measured using Eq. 3.1, and the conflict pixels are assigned to the lowest fitness value cluster among those adjacent clusters (Siddiqui & Isa, 2012). This makes sure that a low variance cluster will move to the active region during the clustering process.

$$f(c_j) = \sum_{i \in c_j} (p_i(x,y) - c_j)^2 \quad (3.1)$$

In case the adjacent clusters of conflict pixels are empty, zero variance, and have positive variance, the pixels-assigning process will measure different analysis to ensure conflict pixels will assign to empty cluster. Thus, the empty center and trapped centroid problem will not occur in the next iteration of the clustering process. In detail, the intensity of conflict pixels will be arranged in ascending order based on their distance from the cluster with the highest fitness value. The intensity of pixels is stored in an array, and it is denoted as E_g, where g=1, 2.... ($k-1$). If the image is segmented into k number of clusters, the k-1 will be the maximum intensity level. Empty cluster and zero variance cluster are difficult to differentiate by k-means clustering. Therefore, the OKM clustering considered the fitness value and the number of pixels of each adjacent cluster. First, all the clusters are arranged in ascending order based on their fitness values, and it is denoted as the F_q, where q=1, 2....v. The fitness value for each cluster is measured using Eq. 3.2. The zero fitness clusters can be zero variance or empty clusters; therefore, the zero-fitness cluster of the adjacent clusters are rearranged according to the number of pixels present in the clusters. The sorting array F_v is renamed as H_w, where w=1, 2....z (z is the number of clusters with zero variance and no pixels). Finally, the transferring of pixels between the E_g and H_w arrays are started, and the pixels with the intensity value of E_g are transferred to the cluster with the value of H_w. This transferring process between H_w and E_g continues until either the value of w equals z or the value of g is equal to ($k-1$) (Siddiqui & Isa, 2012). In case the transferring process terminates with the value of g equals to z and the value of z is less than k-1, the remaining pixels of E_g will be assigned to the adjacent cluster with the lowest fitness value. After completing the assigning of pixels of E_g, all the centroids are updated using equation 3.2.

$$c_j = \frac{1}{n_j} \sum_{i \in c_j} p_i(x,y) \tag{3.2}$$

The clustering process mentioned above is repeated until the mean square error (MSE) difference is less than α, where $0 < \alpha \leq 1$ (Siddiqui & Isa, 2012). The typical value of α to obtain a good segmentation performance should be close to 0. The terminating criteria could be defined by,

$$MSE^{(t+1)} - MSE^{t} < \alpha \tag{3.3}$$

where, t is the number of iterations.

3.2.1 Implementation of Optimized k-means

Optimized k-means (a.k.a. OKM) clustering technique segment an image I with dimensions $N \times M$ into k number of clusters. The main objective of the OKM clustering technique is to group the image pixels $p_i(x, y)$ into $c_j$clusters. To achieve this objective, it minimizes the cost function, i.e., similar to the cost function of the k-means clustering.

$$OKM = \sum_{i=1}^{N \times M} \min_{j \in \{1,\ldots,k\}} p_i(x,y) - c_j^{\,2} \tag{3.4}$$

According to the study (Siddiqui & Isa, 2012), the optimized k-means clustering is implemented as follows:

```
input: test image with the size of N × M
input: k number of clusters
input: α is a constant value in the range 0 < α ≤ 1,
input: generate a random value of the centroids (c_i)
output: centroids value and segmented image

initialize
    // generate a segmented image with the test image's size, and it contains no information
    on their pixels.
while
for i ← 1 to N do
        for j ← 1 to M do
                // measure the distance between all p_i(x,y) of the test images and centroids.
                // p_i(x,y) with minimum distance assigned to their centroids except those
                pixels that have the same distance to adjacent clusters (i.e. the conflict
                pixels). In other words, labeling the segmented image pixels with the
                centroid value of their nearest clusters.
                // find the conflict pixels.
                // measure the (n_j) number pixels in each cluster.
        end for
end for
// Measure the MSE value for each cluster (it is measured at first iteration only).
// Measure fitness value of each cluster and find the cluster with the highest fitness value
c_l .
// For the conflict pixels, sort the intensity of these pixels in ascending order based on their
distance from c_l and denote the sorting array as E_g, where g=1, 2… (k-1).
// measure pixels in each cluster
if empty cluster is true
        // Sort all clusters in ascending order based on their fitness values and denote the
        sorting array as F_q, where q= 1, 2…v.
        // Sort only zero fitness value clusters in ascending order based on the number of
        pixels in the clusters and denote the sorting array as H_w, where w= 1, 2 …. z.
        // Assign the pixels of E_b to H_b and continue it until the value of b in H_b equals to z
        and the value of b in E_b equals to (k-1). If z < (k-1), then the remaining pixels of E_b
        are assigned to the adjacent cluster with lowest fitness value.
else
        // Assign conflict pixels to lowest fitness value cluster among those adjacent clusters.
end if
// calculate the new value of centroids.
// calculate the MSE value.
if ||MSE^(t+1) − MSE^t|| ≥ α
                // erase all information from the segmented image.
                // c_i = c_j , erase all information from c_j .
else
        // return the segmented image and information centroid (c_j)
        // terminate all loop immediately
end if
end while
```

The flow diagram of the OKM clustering technique is shown in Fig. 3.1. Comparatively, k-means clustering has a short processing time, and its time complexity is defined as (Leibe et al., 2006):

$$timecomplexity_{k-means} = O(ndkt) \tag{3.5}$$

The time complexity of k-means have parameters, such as n is the number of pixels in an image, d is the number of attributes' dimensions, k is the number of clusters, and t is the number of iteration.

The OKM clustering is based on the k-means and has an additional transferring process, so the time complexity of the OKM is defined as:

$$timecomplexity_{OKM} = O(ndktb) \tag{3.6}$$

The only additional parameter in the time complexity equation of the OKM as compared to the k-means clustering is b, which is the intensity values of the conflict pixels that are to be assigned to their respective clusters. At the same time, the big notation Oin both equations defines the growth rate of their function. The time complexity of OKM clustering is not significantly increasing because of the extra parameter b, as the advanced process involves only implementing if the conflict pixels and empty clusters are obtained during the clustering process.

3.2.2 Performance Illustration of Optimized k-means Clustering

The optimized k-means (OKM) clustering's main objective is to avoid empty clusters and trapped centroid problems. To illustrate the OKM performance, the *Bridge* image is selected, as shown in Fig. 3.2 (a). Figure 3.2 (c) and (e) shows segmented images into three and six regions or clusters by the OKM clustering. In addition, the histograms of the test image before and after the segmentation process with different numbers of clusters (i.e., three and six) are shown in Fig. 3.2 (b), (d), and (f), respectively. Comparing histograms of the test image and OKM segmented image with three and six number of clusters confirms that OKM produced the final centroids lactation at the appropriate peak's original histogram as shown in Fig. 3.2 (d) and (f), and their final solutions are converged to the nearby global optimum location. In Fig. 3.2 (c) and (f) , the possible region of objects like steel ropes, deck, pillars, mountains, and sky are significantly segmented. These results show that the OKM can avoid the centroid being trapped in the non-active region. In addition, both segmented images have no dead center, and both images are clustered into three and six regions, respectively. Thus, the results confirm that the OKM clustering is also capable of avoiding the empty cluster problem. Table 3.1 tabulated the values of MSE, INTER, F(I), F′ (I), and Q(I) functions, where the small value of the

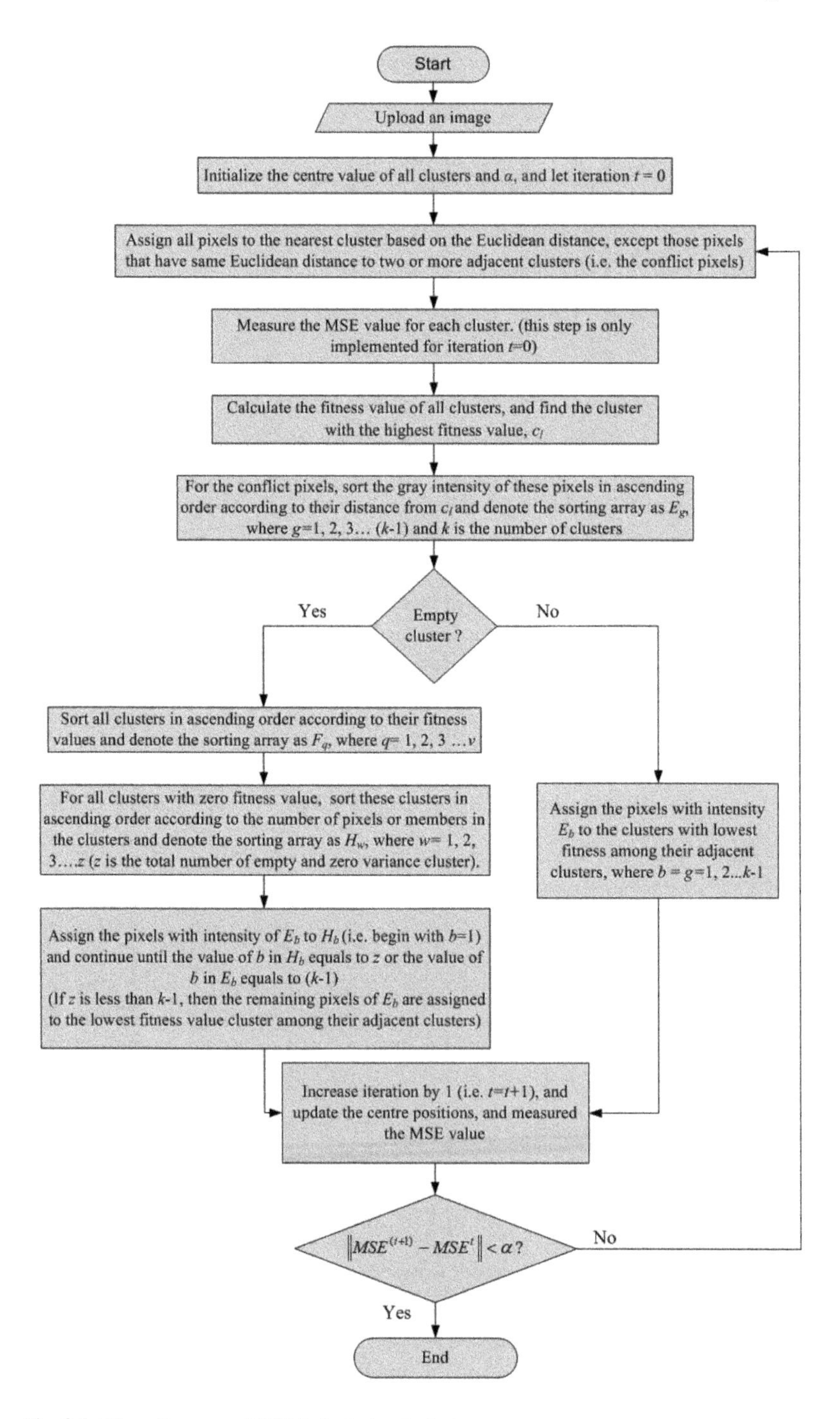

Fig. 3.1 Flow diagram of OKM clustering technique

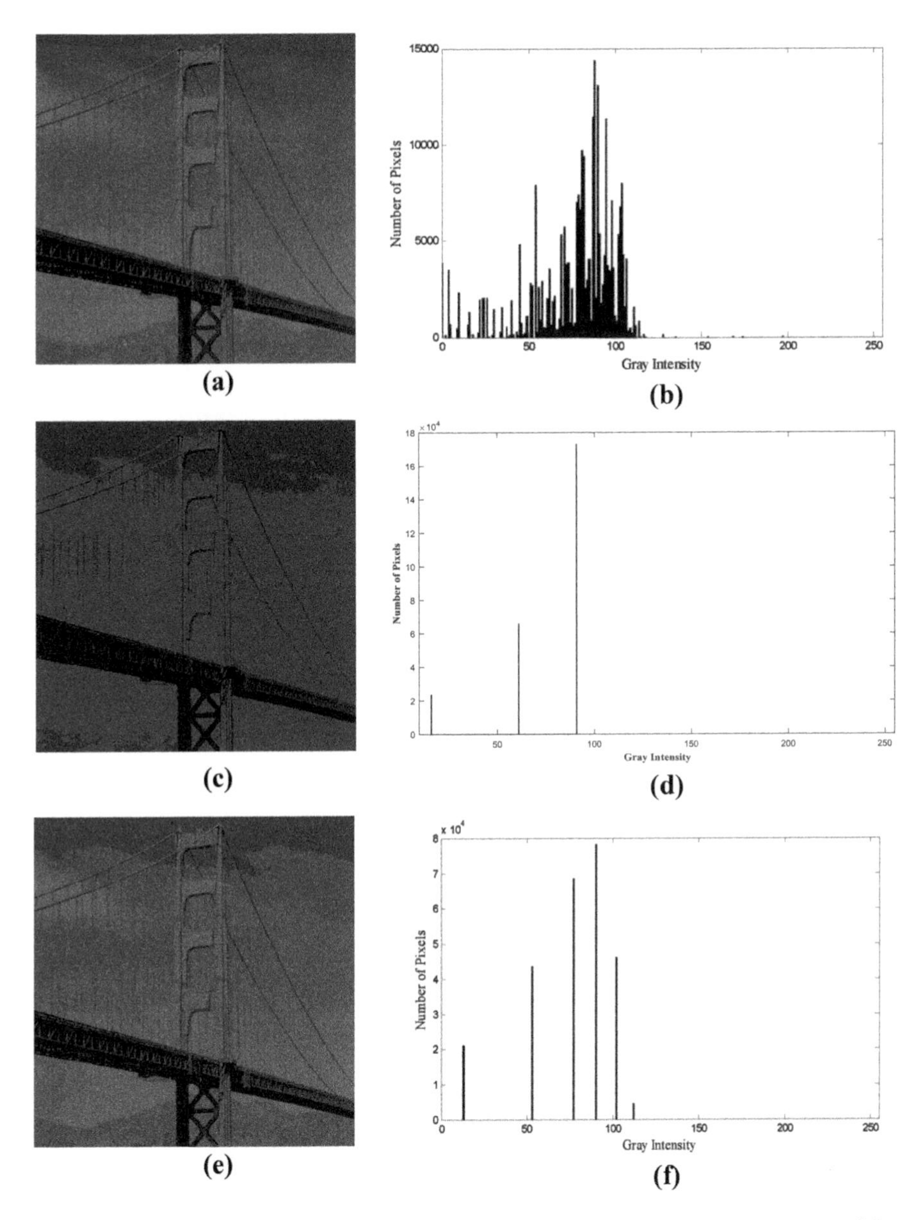

Fig. 3.2 Applying the OKM clustering on ***Bridge*** image; (**a**) original image, (**b**) histogram of the original image, (**c**) segmented image at k=3, (**d**) histogram of the segmented image at k=3, (**e**) segmented image at k=6, and (**f**) histogram of the segmented image at k=6

MSE and large value of the INTER confirm that images are homogeneously segmented and their solutions are converged to the optimum global location. Furthermore, the small values of F(I), F′ (I), and Q(I) also confirm that the OKM clustering is successful in segmenting the images into homogenous regions.

Table 3.1 Results produced by the OKM clustering technique on ***Bridge*** and Light house images

Quantitative test	Bridge	Light house
Initialized number of clusters	6	3
Final number of clusters	6	3
MSE	30.382	92.623
INTER	43.666	49.14
$F(I)$ 1x 10^7	5.198	3.26
$F'(I)$	7.324	2.304
$Q(I)$	9447.4	85.129

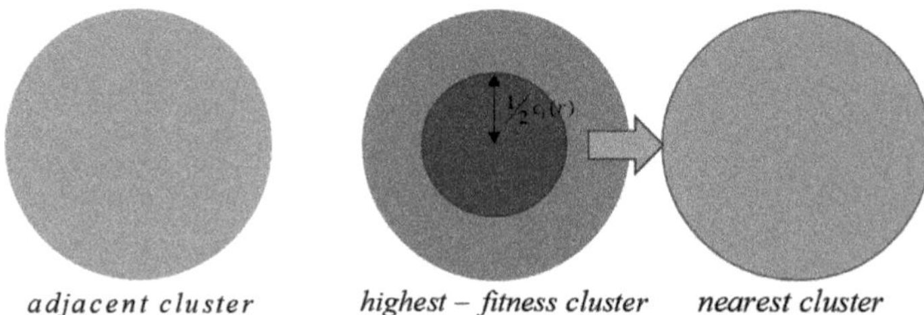

Fig. 3.3 Transferring process of EMKM1 for the cluster with the highest fitness value, where members of c_l outside the range (shows in light blue region) are transferred to the nearest cluster only

3.3 Enhanced Moving k-means Clustering

The enhanced moving k-means (EMKM) clustering was introduced by (Siddiqui & Isa, 2011); it employs a hard membership function, where the pixels are assigned to a single cluster only. After assigning all pixels to clusters, the positions of centroids are updated in each iteration of the clustering process according to Eq. 3.2. The EMKM clustering technique measures each cluster's fitness in the same fashion used in the conventional MKM and AMKM clustering techniques. The clusters are arranged in ascending order where the cluster with the smallest fitness values and the cluster with the largest fitness value is denoted as c_s and c_l, respectively. The relationship between c_s and c_l must fulfil the condition $f(c_s) \geq \alpha_a f(c_l)$ to secure a final solution at the optimum location. The variance among clusters is not considered similar if the condition $f(c_s) \geq \alpha_a f(c_l)$ is not fulfilled. However, the MKM and AMKM clustering techniques have an inappropriate transferring process that failed to keep a variance balance among the cluster.

Thus, the two versions called EMKM1 and EMKM2 were introduced in the literature to keep the cluster variance in a reasonable range and avoid the cluster trap at an optimum local location. The EMKM1 clustering technique c_l will keep its members within the range of $\frac{1}{2}c_l(r)$ where $c_l(r)$ is the radius of the cluster c_l (Siddiqui & Isa, 2011). As shown in Fig. 3.3, the members with a value more than the defined threshold $\frac{1}{2}c_l(r)$ will be assigned to the nearest cluster. The cluster's variance c_l is reduced significantly because of transferring the cluster's outer pixels with the highest fitness value. Overall, this process ensures a balance of fitness value between clusters with the highest fitness value and other clusters by significantly reducing each cluster variance. In addition, if pixels of the nearest cluster are outside the range of $\frac{1}{2}c_{nc}(r)$, where $c_{nc}(r)$ is the radius of the cluster c_{nc} and c_{nc} is nearest cluster to the cluster with the smallest fitness value, then these pixels will be transferred to the cluster with the smallest fitness value.

For the EMKM2 clustering technique, the highest fitness value cluster members located at the border of regions c_l and c_j (i.e., where $j = 1, 2, 3, \ldots k; j \neq l$) will be assigned to the cluster c_j. For example, a transferring process is shown in Fig. 3.4, where bordered members of the highest-fitness value are transferred to their closed clusters rather than to only the nearest cluster of the highest-fitness value cluster. Figure 3.4 shows the two border regions of the highest fitness value cluster, which contains transferring members, are highlighted by yellow and light blue colors. This process is not only capable of reducing the variance of the highest fitness value cluster c_l by giving up border located members, but it also ensures reasonable fitness value among clusters by reducing the summation value of Euclidean distance. In addition, if the bordered members of nearest clusters c_{nc} to the smallest fitness value cluster c_s, these members will be transferred to the cluster c_s.

After completing the transferring process, all centroids and values of α_a fitness condition $f(c_s) \geq \alpha_a f(c_l)$ are updated according to their relation $\alpha_a = \alpha_a - \alpha_a / k$. The mentioned processes repeat themselves until the defined criterion $f(c_s) \geq \alpha_a f(c_l)$ is fulfilled. To ensure a better clustering process, another criterion $f(c_s) \geq \alpha_b f(c_l)$ is employed, and the value α_b is updated according to $\alpha_a = \alpha_o$ and $\alpha_b = \alpha_b - \alpha_b / k$. This process will repeat itself until the defined criterion is not fulfilled.

3.3.1 Implementation of Enhanced Moving k-means Clustering

Let an image I with the dimensions $N \times M$ be clustered into k regions or clusters. Initially, the number of clusters and their centroid values are defined manually. The $p_i(x, y)$ pixels are assigned to their nearest cluster c_j based on the minimum distance,

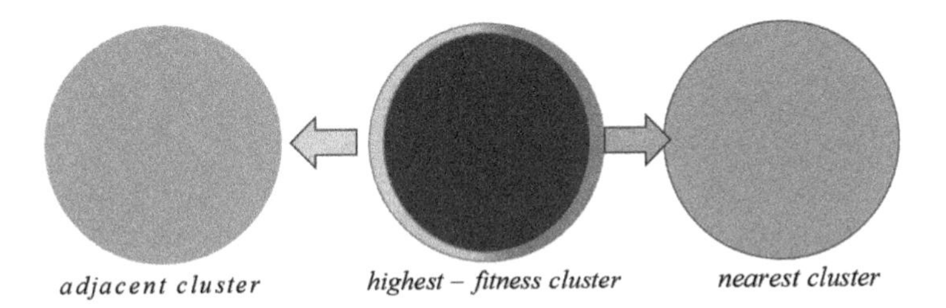

Fig. 3.4 Transferring process of EMKM2 for a cluster with the highest fitness value, where bordered members of the highest fitness value cluster (members located at yellow and light blue regions) are transferred to their closed clusters

where $i = 1, 2 \ldots N \times M$ and $j = 1, 2 \ldots k$. According to the literature (F. U. Siddiqui & Isa, 2011), the EMKM-1 and EMKM-2 clustering techniques are implemented below (Figs. 3.5 and 3.6).

(a) **EMKM-1**

input: test image with the dimensions of $N \times M$
input: k number of clusters
input: generate a random value of the centroids (c_i)
input: set $\alpha_a = \alpha_b = \alpha_o$. (i.e., α_a, α_b and α_o are the small constants, and the α_o value is in the range $0 < \alpha_o < 1/3$).
output: centroids value and segmented image

initialize
// generate a segmented image with the test image's size, and it contains no information on their pixels.
while
  for $i \leftarrow 1$ **to** N **do**
    for $j \leftarrow 1$ **to** M **do**
      // measure the distance between all $p_i(x, y)$ of the test images and centroids.
      // $p_i(x, y)$ with minimum distance assigned to their centroids. In other words, labeling the segmented image pixels with the centroid value of their nearest clusters.
      // measure the (n_j) number pixels in each cluster.
    end for
  end for
  // calculate the value of centroids using Eq. 3.2.
  // measure the fitness value of each cluster using Eq. 3.1.
  // find the cluster with the smallest fitness value c_s and cluster with largest fitness value c_l.
  if $f(c_s) < \alpha_a f(c_l)$ ***/ fitness verification step**
    // analyzed $p_i(x, y)$ of segmented image association with clusters
    if $p_i(x, y) > 1/2\, c_l(r)$, here $i \in c_l$
      // Assign the range of pixels of c_l to the nearest cluster c_{nc} and leave the rest of the pixels in c_l.
    end if
    if $p_i(x, y) > 1/2\, c_{nc}(r)$, here $i \in c_{nc}$
      // Assign the range of pixels of c_{nc} to c_s.
    end if
    // calculate the centroid position of all clusters using Eq. 3.2.
    // update α_a according to $\alpha_a = \alpha_a - \alpha_a / k$
  else
    // measure the distance between all $p_i(x, y)$ of the test images and the updated centroids.
    // $p_i(x, y)$ with minimum distance assigned to their centroids. In other words, labeling the segmented image pixels with the centroid value of their nearest clusters.

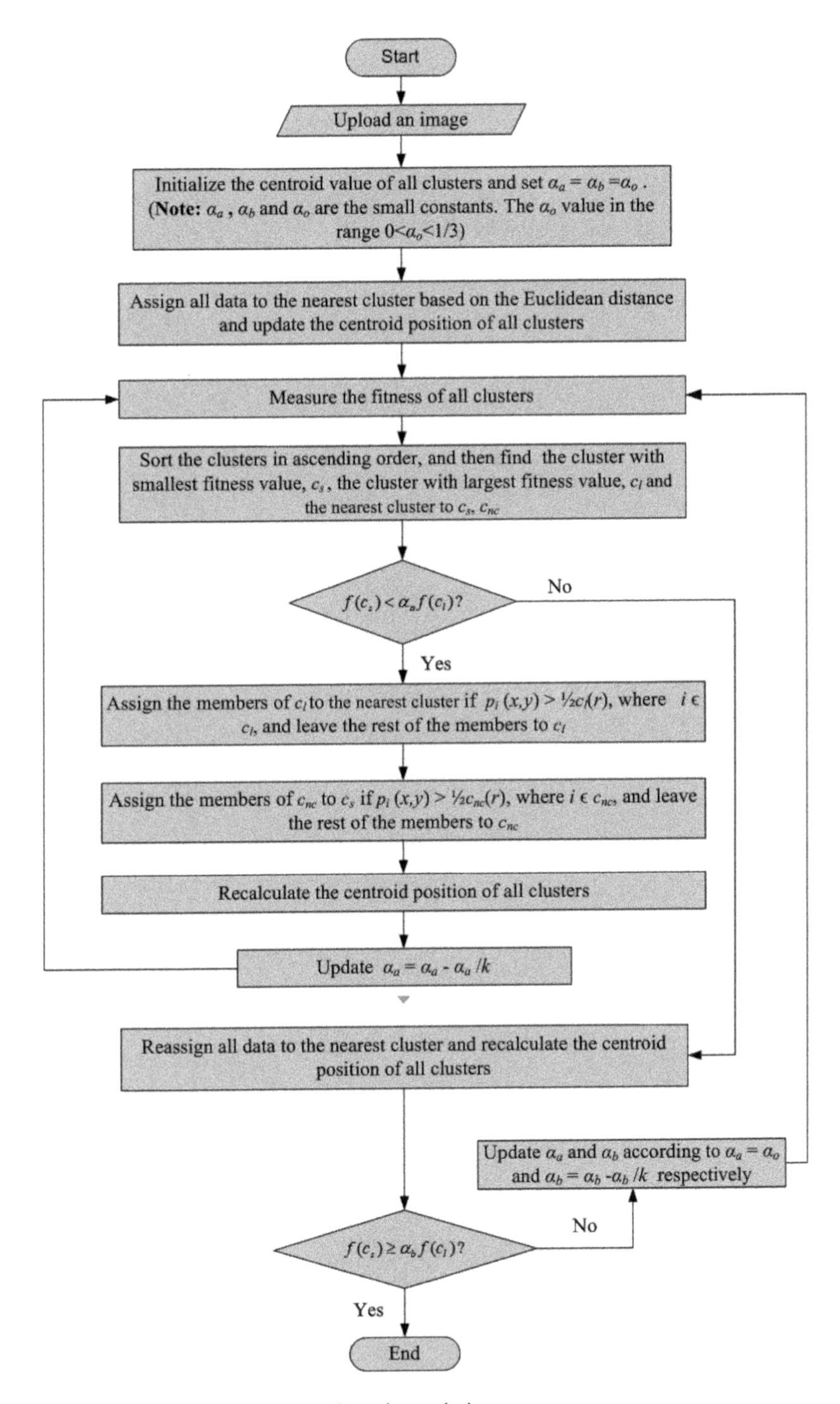

Fig. 3.5 Flow diagram of EMKM-1 clustering technique

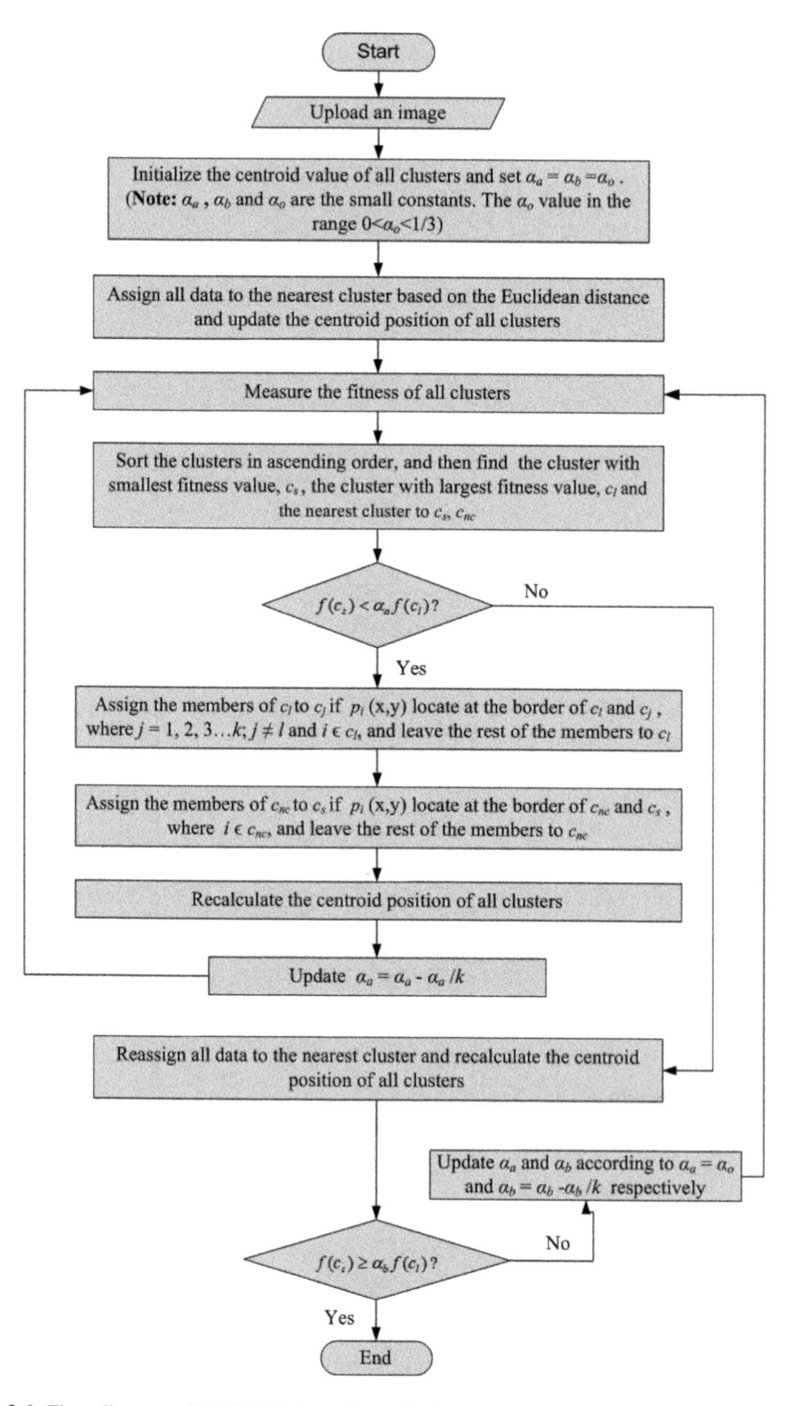

Fig. 3.6 Flow diagram of EMKM-2 clustering technique

(b) **EMKM-2**

```
            // measure the (n_j) number pixels in each cluster.
            // calculate the new value of centroids using Eq. 3.2.
            // measure the fitness value for all clusters using Eq. 3.1.
            if f(c_s) ≥ α_b f(c_l)
                // Update α_a and α_b according to α_a = α_o and α_b = α_b − α_b/k,
                respectively.
                // return to the fitness verification step of this clustering process
            else
                //jump to terminate all process and return the segmented image and
                information centroid (c_j)
            end if
        end if
    end while
```

input: test image with the dimensions of $N \times M$
input: k number of clusters
input: generate a random value of the centroids (c_l)
input: set $\alpha_a = \alpha_b = \alpha_o$. (i.e., α_a, α_b and α_o are the small constants, and the α_o value is in the range $0 < \alpha_o < 1/3$).
output: centroids value and segmented image

initialize
// generate a segmented image with the test image's size, and it contains no information on their pixels.
while
  for $i \leftarrow 1$ **to** N **do**
    for $j \leftarrow 1$ **to** M **do**
      // measure the distance between all $p_i(x, y)$ of the test images and centroids.
      // $p_i(x, y)$ with minimum distance assigned to their centroids. In other words, labeling the segmented image pixels with the centroid value of their nearest clusters.
      // measure the (n_j) number pixels in each cluster.
    end for
  end for
  // calculate the value of centroids using Eq. 3.2.
  // measure the fitness value of each cluster using Eq. 3.1.
  // find the cluster with the smallest fitness value c_s and cluster with largest fitness value c_l.
  if $f(c_s) < \alpha_a f(c_l)$ ***/ fitness verification step**
    // analyzed $p_i(x, y)$ of segmented image association with clusters
    if $p_i(x, y)$ at boundary of the cluster c_l
      // Assign the range of pixels of c_l to the nearest cluster c_{nc} and leave the rest of the pixels in c_l.
    end if
    if $p_i(x, y)$ at the boundary of the cluster c_{nc}
      // Assign the range of pixels of c_{nc} to c_s.
    end if
    // calculate the centroid position of all clusters using Eq. 3.2.
    // update α_a according to $\alpha_a = \alpha_a - \alpha_a / k$
  else
    // measure the distance between all $p_i(x, y)$ of the test images and the updated centroids.

// $p_i(x, y)$ with minimum distance assigned to their centroids. In other words, labeling the segmented image pixels with the centroid value of their nearest clusters.
// measure the (n_j) number pixels in each cluster.
// calculate the new value of centroids using Eq. 3.2.
// measure the fitness value for all clusters using Eq. 3.1.
if $f(c_s) \geq \alpha_b f(c_l)$
// Update α_a and α_b according to $\alpha_a = \alpha_o$ and $\alpha_b = \alpha_b - \alpha_b / k$, respectively.
// return to the fitness verification step of this clustering process
else
//jump to terminate all process and return the segmented image and information centroid (c_j)
end if
end if
end while

3.3.2 *Performance Illustration of Enhanced Moving k-means Clustering*

The enhanced versions of the moving k-means clustering called EMKM-1 and EMKM-2 are capable of avoiding the empty cluster and trapped centroids problems that ensure the final solution converges to the optimum global location. The mentioned superior abilities of EMKM-1 and EMKM-2 Clustering techniques were verified using the *Gantry Crane* image with the number of clusters to four. The segmented images produced by the EMKM-1 and EMKM-2 clustering techniques are shown in Fig. 3.7 (c) and (e), respectively. In addition, the histograms of the original image and segmented images produced by the EMKM-1 and EMKM-2 techniques are also plotted; refer to Fig. 3.7 (d) and (f). In the last chapter, the AMKM has been applied on the same image, and the switch panel is not homogeneously segmented because of the trapped centroid at the non-active region. Whereas EMKM-1 and EMKM-2 homogeneously segment the switch panel, particularly the EMKM-2, which produces remarkable results with homogenous segmentation of the switch panel region. This is also confirmed by the histogram presented in Fig. 3.7 (d) and (f), where the centroid significantly represents an active region with a range of 0–50 on the intensity scale. Furthermore, Table 3.2 presents the selected image's quantitative results for testing the performance of EMKM-1 and EMKM-2. The small MSE value and large INTER value confirm that the final centroids are located at the active regions based on the results. In addition, the small values of the *F(I)*, *F′ (I)*, and *Q(I)* show that EMKM-1 and EMKM-2 techniques segment the image homogenously. In comparison, the EMKM-2 has smaller values of *F(I)*, *F′ (I)*, and *Q(I)* than its variant EMKM-1. This confirms the robust performance of the advanced transferring process of EMKM-2, where only the bordered member of the highest fitness value cluster transfers to their closet clusters of adjacent clusters.

3.4 Outlier Rejection Fuzzy c-means Clustering

Outlier rejection fuzzy c-means (ORFCM) clustering is an advanced version of fuzzy c-means that reduces the outlier sensitivity of the clustering process to converge the final solution at the optimum location (F. U. Siddiqui et al., 2013). Similar to the fuzzy c-means clustering, the ORFCM clustering partially divides the image pixels $I = \{x_i\}$, $i \in \{1, 2, 3 \ldots n\}$ into k the number of overlapping regions or clusters. The main objective of ORFCM is to minimize the cost function,

$$ORFCM = \min \sum_{i=1}^{n} \sum_{j=1}^{k} u_{ij}^{m} \beta^{x_i - c_j^2} \tag{3.7}$$

$$\begin{gathered} subjectto: \\ u_{ij} \in [0,1] fori = 1,2..nj = 1,2..k \\ \sum_{j=1}^{k} u_{ij} = 1, i = 1,2\ldots n \end{gathered}$$

where u_{ij} is the degree of membership of pixelx_i in the jth cluster and m is the degree of fuzziness, i.e., typically equals to 2. Major changes occur in the membership function of the conventional fuzzy c-means clustering to overcome the outlier's sensitivity, i.e., the original Euclidean distance of membership $\|x_i - c_j\|$ is replaced with $(\beta)^{x_i - c_j}$. According to (F. U. Siddiqui et al., 2013), the modified equation in calculating membership u_{ij} is given by:

$$u_{ij} = \frac{1}{\sum_{p=1}^{k} \left[\frac{\beta^{x_i - c_j}}{\beta^{x_i - c_p}} \right]^{(2/m-1)}} \tag{3.8}$$

The exponent variable β limits the partial distribution of points among the two neighboring clusters rather than to all clusters, and it is defined as:

$$\beta = \frac{(RangeofIntensityinanimage) + 1}{MaximumRangeofIntensity + 1} + 1 \tag{3.9}$$

In case of gray images, the above equation can be defined as:

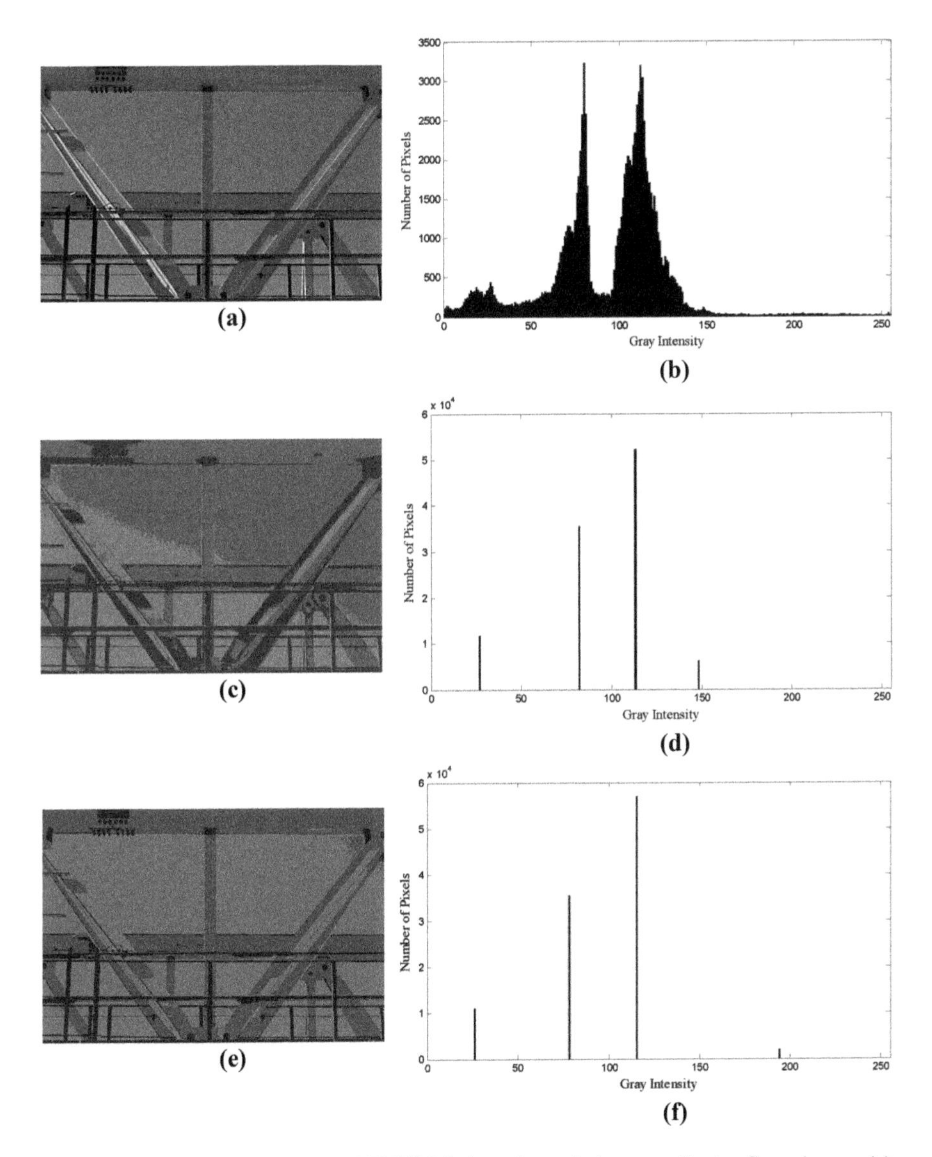

Fig. 3.7 Applying the EMKM-1 and EMKM-2 clustering techniques on ***Gantry Crane*** image; (**a**) original image, (**b**) histogram of the original image, (**c**) EMKM-1 segmented image, (**d**) histogram of EMKM-1 segmented image, (**e**) EMKM-2 segmented image, and (**f**) histogram of EMKM-2 segmented image

Table 3.2 Results produced by the EMKM-1 and EMKM-2 clustering techniques on ***Gantry Crane*** image

Quantitative Test	EMKM-1	EMKM-2
Initialized number of clusters	4	4
Final number of clusters	4	4
MSE	160.621	118.647
INTER	65.66	90.16
F(I) 1× 10^7	4.393	2.769
F'(I)	5.418	2.601
Q(I)	27.256	7.5949

$$\beta = \frac{(I_{max} - I_{min}) + 1}{256} + 1 \tag{3.10}$$

where, the I_{max} is the maximum intensity in an image and I_{min} is the minimum intensity in the gray image. The value of β is between 1 and 2, it is close to 2 if the image intensity range is maximum. The membership with $\beta = 2$ can restrict the partial distribution of points between two adjacent clusters. Whereas, membership with $\beta = 1$ (i.e., occur if image intensity range is lowest) becomes more flexible only partially to distribute the adjacent clusters' points.

3.4.1 Implementation of Outlier Rejection Fuzzy c-means Clustering

Let I denote an image with n pixels (i.e., $p_i(x, y)$ $i \in \{1, 2, \ldots . n\}$) to be partitioned into k clusters, where $2 \leq k \leq n$ and c_j (for j=1,2,….k) be the jth cluster. Consider the matrix $U = (u_{ij})_{k \times n}$ called fuzzy partition matrix in which each element u_{ij} indicates the membership degree of each pixel in the j^{th}cluster, c_j. The implementation of ORFCM are as follows:

input: test image with the size of $N \times M$
input: define the value of termination criteria (set within the range of 0 and 1 and here $\varepsilon = 0.001$) and degree of fuzziness (e.g., $m = 2$)
input: k number of clusters
input: randomly generate membership function u_{ij} for all clusters.
output: centroids value and segmented image

initialize
// generate a segmented image with the test image's size, and it contains no information on their pixels.
while
for $i \leftarrow 1$ **to** N **do**
for $j \leftarrow 1$ **to** M **do**
// calculate the cluster centroid c_j according to Eq. 3.11.

$$c_j = \frac{\sum_{i=1}^{n} \left(u_{ij}^{m} \times p_i(x, y)\right)}{\sum_{i=1}^{n} u_{ij}^{m}} \quad \textbf{(3.9)}$$

// $p_i(x, y)$ with the highest membership are assigned to their respective centroids. In other words, the segmented image pixels are labeled with the centroid value of their clusters.
// measure the distance between all $p_i(x, y)$ of the test images and centroids.
//compute the new membership function u_{ij} based on Eq. 3.12.

$$u_{ij} = \frac{1}{\sum_{p=1}^{k} \left[\frac{\beta^{\|x_i - c_j\|}}{\beta^{\|x_i - c_p\|}} \right]^{(2/m-1)}} \quad \textbf{(3.10)}$$

end for
end for
if $\left\|u_{ij}^{(t+1)} - u_{ij}^{t}\right\| < \varepsilon$
// terminate the process.
end if
// erase all information from the segmented image and c_j.
end while

The flow diagram of the outlier rejection fuzzy c-means (ORFCM) clustering is shown in Fig. 3.8.

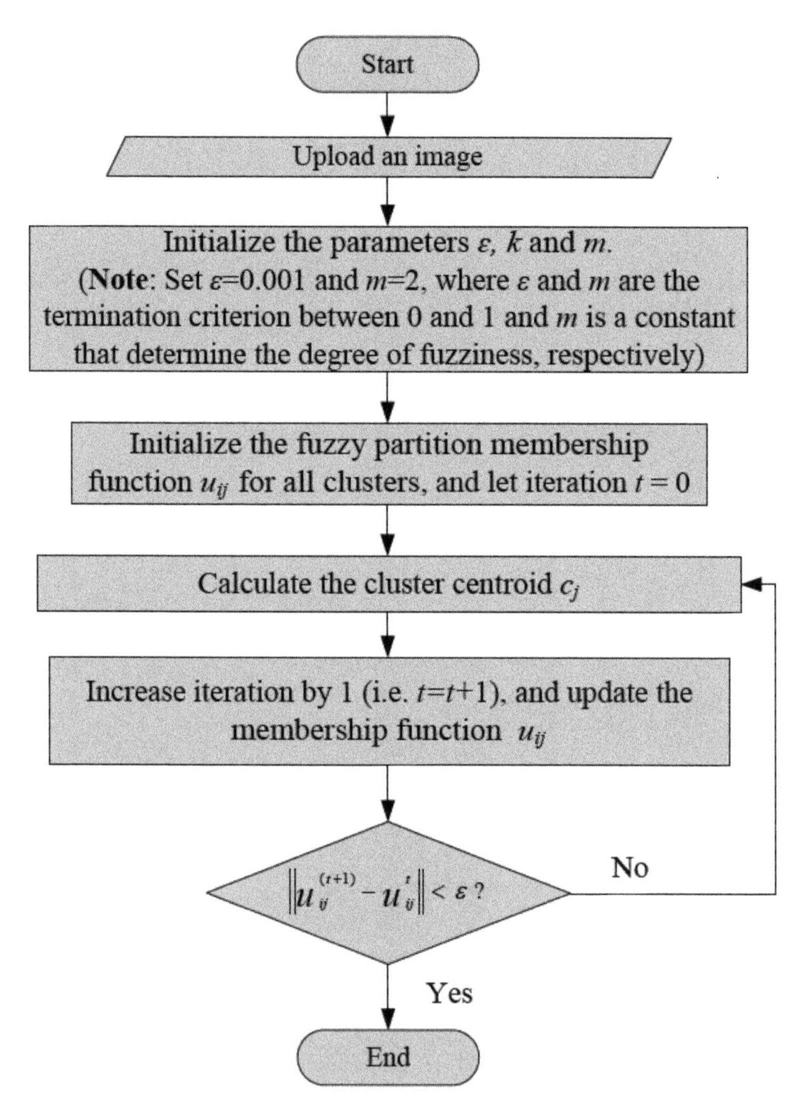

Fig. 3.8 Flow diagram of outlier rejection fuzzy c-means clustering

3.4.2 Performance Illustration of Outlier Rejection fuzzy c-means Clustering

To illustrate the ORFCM clustering technique's performance, two data ranges were employed, i.e., 1 to 120 and 1 to 60. After applying the ORFCM technique on both data, the final obtained regions c_1, c_2, and c_3 with their respective membership functions u_1, u_2, and u_3 are plotted in Fig. 3.9 (a) and (b). In the graph, all the data are partially distributed between only the two clusters. The graph also confirms that outliers are not effective in the clustering process; for example, the graph's outliers have almost zero membership value. More importantly, the data of large intensity range (i.e., 1–120) is partial distribution between the two adjacent clusters, and the data of small intensity range (i.e., 1–60) is not only partially distributed between two adjacent clusters but their member between the two adjacent clusters is much flexible. Thus, the effects of outliers in the clustering process have been successfully

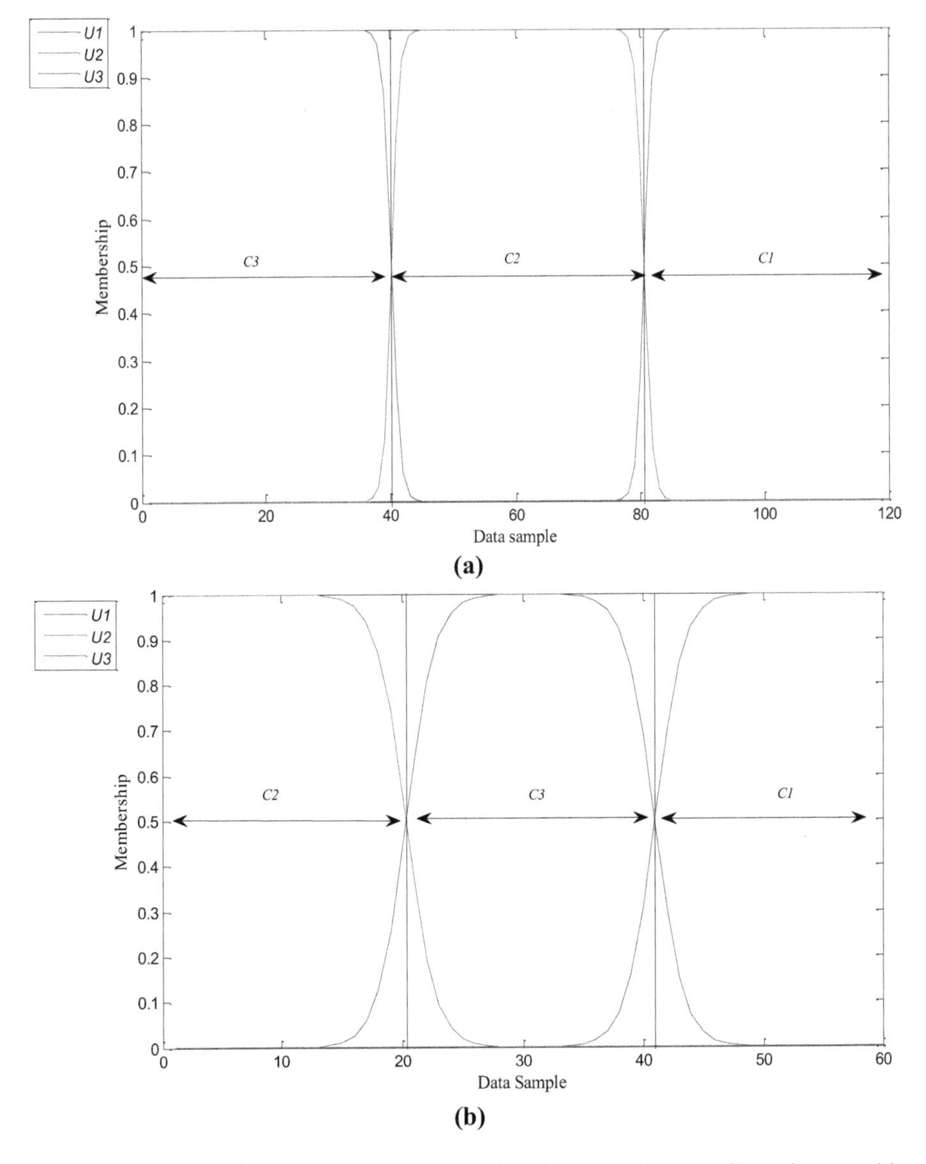

Fig. 3.9 Membership function generated by the ORFCM clustering for data of intensity range (**a**) 1–120 and (**b**) 1–60. (Siddiqui et al., 2013)

reduced, reducing the chance of the center trapping problem of small intensity range data.

The *Football* image is also selected further to evaluate the performance of the ORFCM Clustering technique. In the initial setting, three clusters are set. As shown in Fig. 3.10 (d), the ORFCM segments the *Football* image into predefined three clusters with accuracy such that the final centroids positioned at 50, 140, and 245 levels in the intensity histogram. By avoiding the outlier effect on the clustering process, regions like the laces of football are homogenously segmented, as shown in Fig. 3.10. The success of ORFCM in the obtained result is further confirmed by quantitative analysis. The small values of the MSE and VXB and the large value of the INTER indicate the better performance of ORFCM (as tabulated in Table 3.3). In addition, the small values of *F(I)*, *F′ (I)*, and *Q(I)* functions confirm homogenous segmentation of regions in the image by the ORFCM clustering technique.

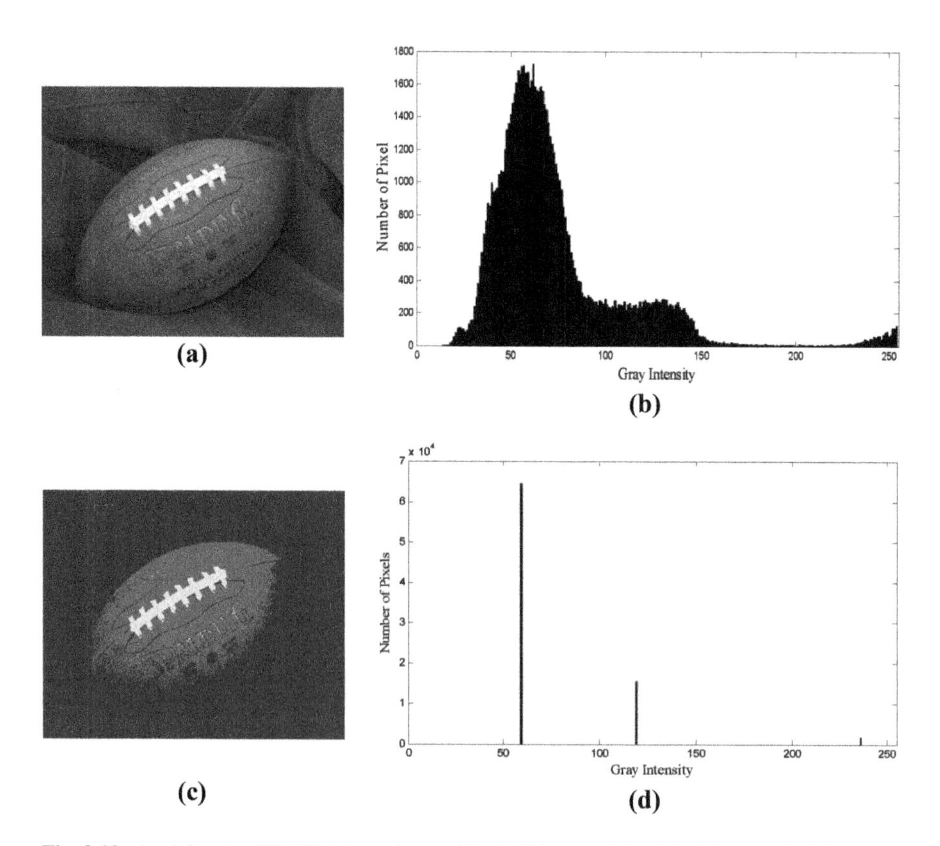

Fig. 3.10 Applying the ORFCM clustering on ***Football*** image; (**a**) original image, (**b**) histogram of the original image, (**c**) segmented image, and (**d**) histogram of the segmented image

Table 3.3 Results produced by the ORFCM Clustering technique on ***Football*** image

Quantitative test	ORFCM
Initialized number of clusters	3
Final number of clusters	3
MSE	230.77
INTER	118
VXB	0.2068
$F(I)$ 1×10^7	1.91558
$F'(I)$	4.094
$Q(I)$	32.9795

References

Leibe, B., Mikolajczyk, K., & Schiele, B. (2006). Efficient clustering and matching for object class recognition. In BMVC, pp. 789–798.

Siddiqui, F. U., & Isa, N. A. M. (2011). Enhanced moving K-means (EMKM) algorithm for image segmentation. *IEEE Transactions on Consumer Electronics, 57*(2), 833–841.

Siddiqui, F., & Isa, N. M. (2012). Optimized K-means (OKM) clustering algorithm for image segmentation. *Opto-Electronics Review, 20*(3), 216–225.

Siddiqui, F. U., Isa, N. A. M., & Yahya, A. (2013). Outlier rejection fuzzy c-means (ORFCM) algorithm for image segmentation. *Turkish Journal of Electrical Engineering & Computer Sciences, 21*(6), 1801.

Chapter 4
Quantitative Analysis Methods of Clustering Techniques

4.1 Analysis Methods of Clustering Techniques

Currently, two analysis methods are available for comparing the performance of the clustering techniques, i.e., qualitative and quantitative analyses.

4.1.1 Qualitative Analysis

Qualitative analysis is the traditional procedure to evaluate the segmentation performance of clustering techniques using human visual perception. An examiner ranks the segmented results based on the following criteria:

- the homogeneity in the segmented regions
- shape information preserves in the segmented image
- quality of edges at the segmented regions
- contrast between the segmented regions

However, it is an extremely laborious job and needs an expert examiner to conduct this analysis. Furthermore, a fair comparison of clustering techniques is a big challenge in qualitative analysis, mainly when the image is segmented into a higher number of clusters. For example, the resultant images of k-means and EMKM-1 clustering techniques are qualitatively analyzed in Fig. 4.1 and Fig. 4.2, where the resultant images are segmented into 3, 4, 5, and 6 clusters. The images of the publicly available dataset/benchmark are using in this book for performance evaluation (Martin et al., 2001). Based on an examiner, the segmented images' key findings are marked by arrows highlighting the major differences between them.

Based on the highlighted marks in Fig. 4.1, the EMKM-1 clustering technique segments the *building* and *plane* images with higher homogeneity in contrast to the

F.U. Siddiqui, A. Yahya, *Clustering Techniques for Image Segmentation*,
https://doi.org/10.1007/978-3-030-81230-0_4

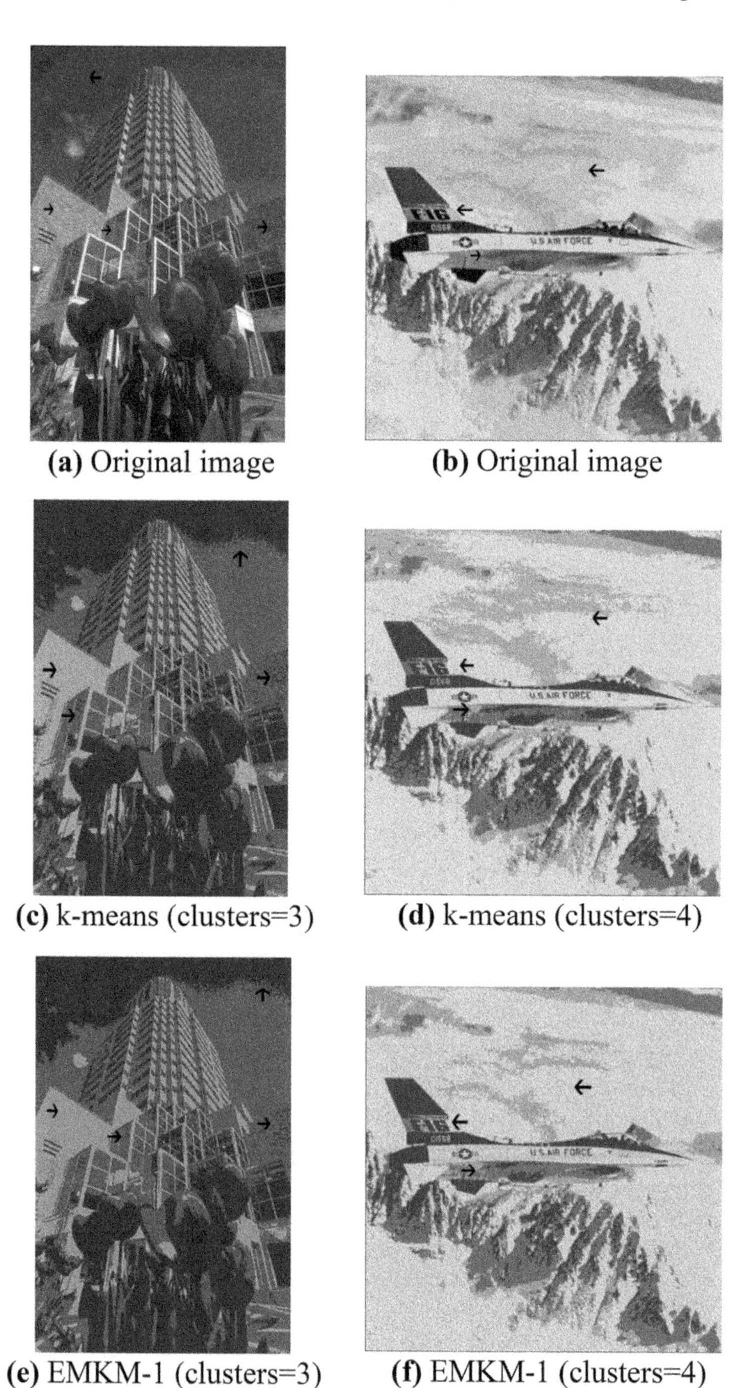

Fig. 4.1 Segmented images in 3 and 4 clusters for the *building* and *plane* images after applying k-means and EMKM-1 clustering techniques. (Fasahat U Siddiqui, 2012)

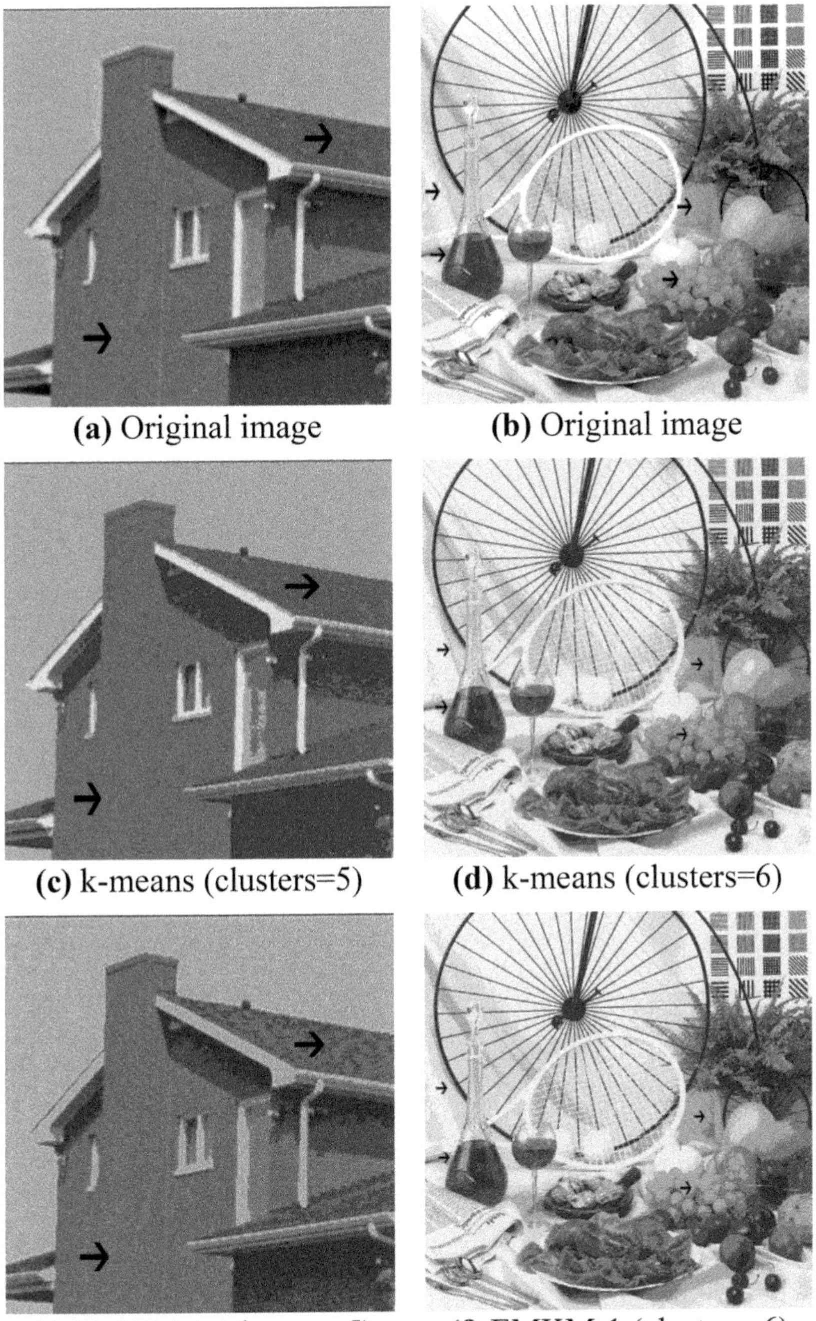

(a) Original image

(b) Original image

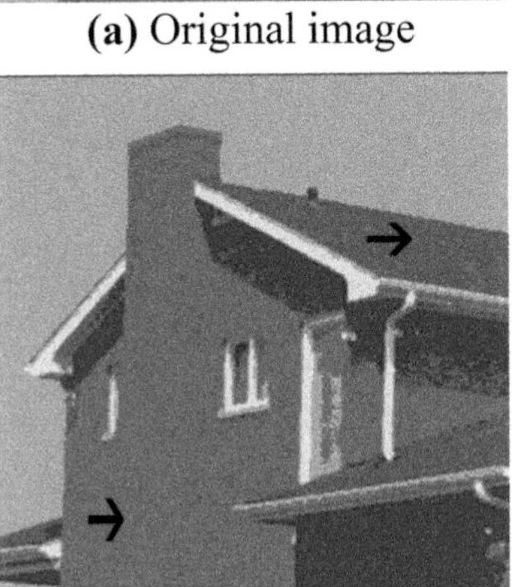

(c) k-means (clusters=5)

(d) k-means (clusters=6)

(e) EMKM-1 (clusters=5)

(f) EMKM-1 (clusters=6)

Fig. 4.2 Segmented images in 5 and 6 clusters for the *House* and *Fruit-table* images after applying k-means and EMKM-1 clustering techniques. (Fasahat U Siddiqui, 2012)

k-means clustering technique. Furthermore, it is impossible to analyze the shapes and edges details of extracted regions by clustering techniques at the lower number of clusters. In terms of the contrast between the segmented regions, both clustering techniques show good results.

In Fig. 4.2, the EMKM-1 clustering segments the *house* and *fruit-table* images with sharp edges and more content than k-means clustering. However, the homogeneity of the extracted regions by the EMKM-1 clustering technique is not the best. Because fewer criteria favor the clustering technique in segmentation of the image into a higher number of clusters, the qualitative analysis becomes difficult even for an expert examiner.

4.1.2 *Quantitative Analysis*

Quantitative analysis is another procedure that can examine the performance of the clustering techniques. It measures two characteristics of the extracted regions, i.e., similarity index and homogeneity of extracted regions. As compared to qualitative analysis, quantitative analysis has many advantages, like it has no human dependency. Therefore, a quick and fair comparison is possible for image segmentation with any number of clusters.

4.1.2.1 Mean Square Error (MSE)

The real-world images' regions have high texture and extremely irregular boundaries. In such a complex environment, the similarity index measurement evaluates the quality without analysis of contrast, shape, homogeneity, and edges. MSE (mean square error) is the simplest similarity index-based quantitative method that measures the square error between the pixels of original and resultant images using Eq. 4.1.

$$MSE = \frac{1}{n}\sum_{j=1}^{k}\sum_{i \in c_j}\left(p_i\left(x,y\right)-c_j\right)^2 \tag{4.1}$$

Where c_j is the jth cluster and $p_i(x, y)$ is the pixel that belongs to j^{th} cluster. It is implemented as follow:

```
input: segmented image
input: original image
output: MSE value

initialize
   // measure image size, i.e., N × M , and number of clusters k
for i ← 1 to N do
      for j ← 1 to M do
                  // measure the difference between all p_i(x, y) of the original image
                  and c_j of the segmented image.
      end for
end for
   // Sum all the difference value of pixels
   // Measure the (n) number pixels in an image.
   // Measure the MSE value and terminate the process
```

Table 4.1 Average MSE value for 103 benchmark images after applying different clustering techniques

Techniques	For k=3	For k=4	For k=5	For k=6
KM	303.52	181.39	115.68	84.715
FCM	297.45	171.66	109.74	**76.746**
MKM	494.59	316.83	229.63	173.64
AMKM	337.94	192.98	131.91	91.622
AFMKM	950.30	456.17	260.01	96.242
AFKM	3745.3	3542.6	3543.1	3593.0
OKM	300.06	179.12	121.72	85.388
EMKM-1	321.26	185.07	119.62	83.853
EMKM-2	315.59	180.08	115.31	80.734
ORFCM	**294.88**	**169.42**	**108.81**	77.310

According to Eq. 4.1, the lower difference between the resultant and original images produces the smaller value of MSE. This indicates that all the pixels in an image are clustered to their nearest clusters such that each cluster has the most similar intensity value pixels. Furthermore, the smaller value of MSE confirms that the final solution is converged to the optimum global location. For example, the average value of MSE for 103 benchmark images is tabulated in Table 4.1. The smallest average value of MSE shows the best performance of a clustering technique at the particular number of clusters (Table 4.2).

Table 4.2 Average INTER value for 103 benchmarks images after applying the clustering techniques.

Techniques	k=3	k=4	k=5	k=6
KM	85.106	83.197	83.197	80.888
FCM	83.585	80.402	80.402	75.845
MKM	71.734	68.111	68.111	66.978
AMKM	84.135	81.190	81.190	79.605
AFMKM	76.530	**84.817***	**84.817***	**88.143***
AFKM	56.161	46.647	43.471	41.513
OKM	**85.650**	83.051	83.051	79.144
EMKM-1	78.679	76.079	76.079	72.746
EMKM-2	82.990	80.490	80.490	82.990
ORFCM	83.074	80.694	80.694	75.290

4.1.2.2 INTER

The similarity index is measured by calculating the variance among the clusters, which distinguishes the differences between adjacent clusters. It is measured by using the following equation.

$$INTER = mean_{\forall r \neq q}\left(c_r - c_q^{\,2}\right) \tag{4.2}$$

Where $r=1, 2,...,(k\text{-}1)$, and $q= (r+1),..., k$.

Here, the inter-cluster variance is measured by taking a mean of difference among the clusters' centroid. The INTER is implemented as follow:

input: segmented image
input: original image
output: Inter value

initialize
// Segmented image pixels are labelled with their centroid value; therefore, the unique pixel value gives the centroid value and number of centroids.
// Sum all the difference value of centroids
// Measure the number of centroids in an image
// Measure the inter value and terminate the process

The large value of INTER shows that the grouped data in the clusters are significantly different from the other clusters and the obtained centroids converged to their optimum locations. Unlike MSE that produced linear output for all situations, the INTER function produces a higher value if dead centroid intensity is zero and low if the dead centroid intensity is similar to the other cluster centroid intensity. As a higher number of clusters favors the dead centroid occurrence, the INTER is not the best quantitative analysis for images segmented into more than two clusters.

*The largest INTER value for the clustering techniques that fail to segment the images into defined number of clusters

4.1.2.3 VXB

Xie and Beni (1991) introduced the VXB function (Pakhira et al., 2004; Xie & Beni, 1991). The VXB is mainly used to analyze the outlier sensitivity of fuzzy-based clustering techniques, which cannot be analyzed by applying the MSE and INTER quantitative analyses. The VXB function measures the compactness and separation of the pixels clustered by the fuzzy-based clustering techniques. It is defined as:

$$VXB = \frac{\sum_{i}^{n}\sum_{j}^{k} u_{ij}^{2} p_i(x,y) - c_j^{2}}{n\left(\min_{\forall r \neq q} c_q - c_r^{2}\right)} \tag{4.3}$$

where u_{ij} is the fuzzy membership of pixel i belongs to j-th cluster. The VBX is implemented as follow:

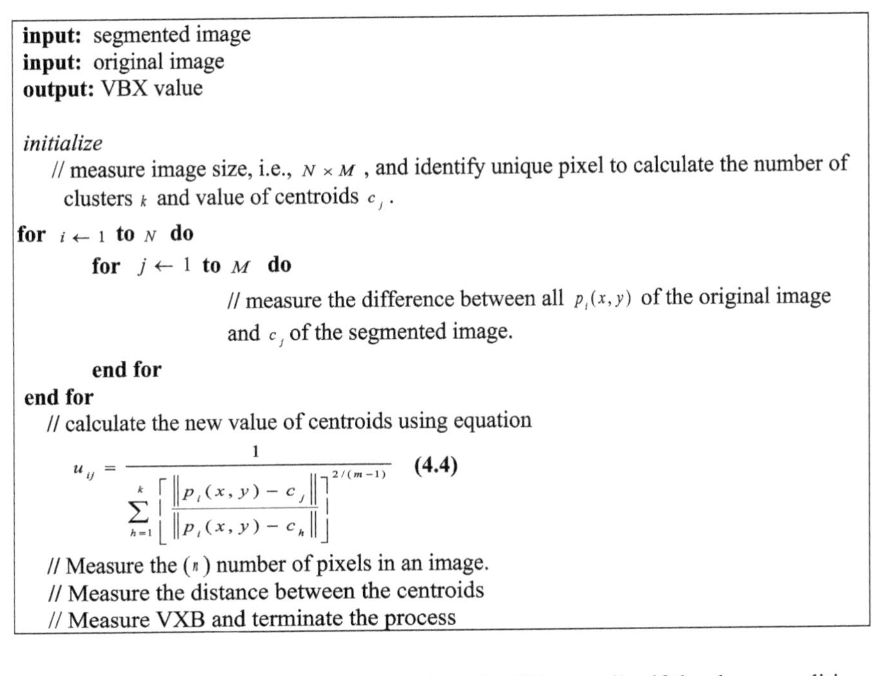

input: segmented image
input: original image
output: VBX value

initialize
// measure image size, i.e., $N \times M$, and identify unique pixel to calculate the number of clusters k and value of centroids c_j.
for $i \leftarrow 1$ **to** N **do**
for $j \leftarrow 1$ **to** M **do**
// measure the difference between all $p_i(x,y)$ of the original image and c_j of the segmented image.
end for
end for
// calculate the new value of centroids using equation

$$u_{ij} = \frac{1}{\sum_{h=1}^{k}\left[\frac{\left\|p_i(x,y) - c_j\right\|}{\left\|p_i(x,y) - c_h\right\|}\right]^{2/(m-1)}} \quad \textbf{(4.4)}$$

// Measure the (n) number of pixels in an image.
// Measure the distance between the centroids
// Measure VXB and terminate the process

The ratio of compactness and separation of will be smaller if the three conditions become true:

- clusters are comparatively less overlapping
- the pixels within the cluster are similar as possible
- the pixels among the cluster are dissimilar as possible

Based on the tabulated information in Table 4.3, AFKM shows infinity value because of dead centroid generation in the clustering process. The best value of

Table 4.3 Average VXB value for 103 benchmark images after applying the fuzzy-based clustering techniques

Techniques	VXB value for different number of clusters			
	k=3	k=4	k=5	k=6
FCM	0.3349	0.3862	0.4199	0.4617
AFMKM	2.4320	2.8931	1.8536	0.8685
AFKM	∞	∞	∞	∞
ORFCM	**0.2593**	**0.2797**	**0.2921**	**0.3052**

VXB is bolded in the table to highlighted them. Based on the results, ORFCM produces lower overlapped clusters; the technique is less sensitive to the outliers than other fuzzy-based clustering techniques.

4.1.2.4 F(I)

In 1994, a novel function was introduced called *F(I)* (Liu & Yang, 1994). It measures the homogeneity of the segmented image without any human interaction or any predefined threshold value. In addition, it indirectly measures the shape and edge information of segmented regions. Furthermore, it also measures the similarity index of the segmented regions. In detail, the function *F(I)* is considering three basic conditions, which are as follows:

- the regions should be uniform and homogenous
- the region's interiors have uniform characteristics or without many holes inside it
- adjacent regions must present significantly different values

The evaluation function is defined as:

$$F(I)=\sqrt{R}\sum_{i=1}^{R}\frac{e_i^2}{\sqrt{A_i}} \tag{4.5}$$

where I is the image to be segmented, R is the number of regions found in the segmented image, A_i is the size of the segmented region, and e_i is the average intensity error and defined as the sum of Euclidean distance of the intensity between the pixels of segmented regions and the original image. In equation 4, two terms mainly penalize the segmentation if the segmented image is not fulfilling the three conditions mentioned above. For instance, the term $\sqrt{R}$ penalizes the segmentation that forms too many regions. Whereas the term $e_i^2/\sqrt{A_i}$ penalizes the small size regions with the large intensity error. The smaller value of *F(I)* indicates a better segmentation result. Based on the experiment conducted in (Liu & Yang, 1994), many regions in the segmented image penalized the segmentation only by the former term. Moreover, the F*(I)* function favors the noisy segmentation as the e_i^2 is often close to zero for small and lower intensity error regions (Borsotti et al., 1998). The *F(I)* is implemented as follow:

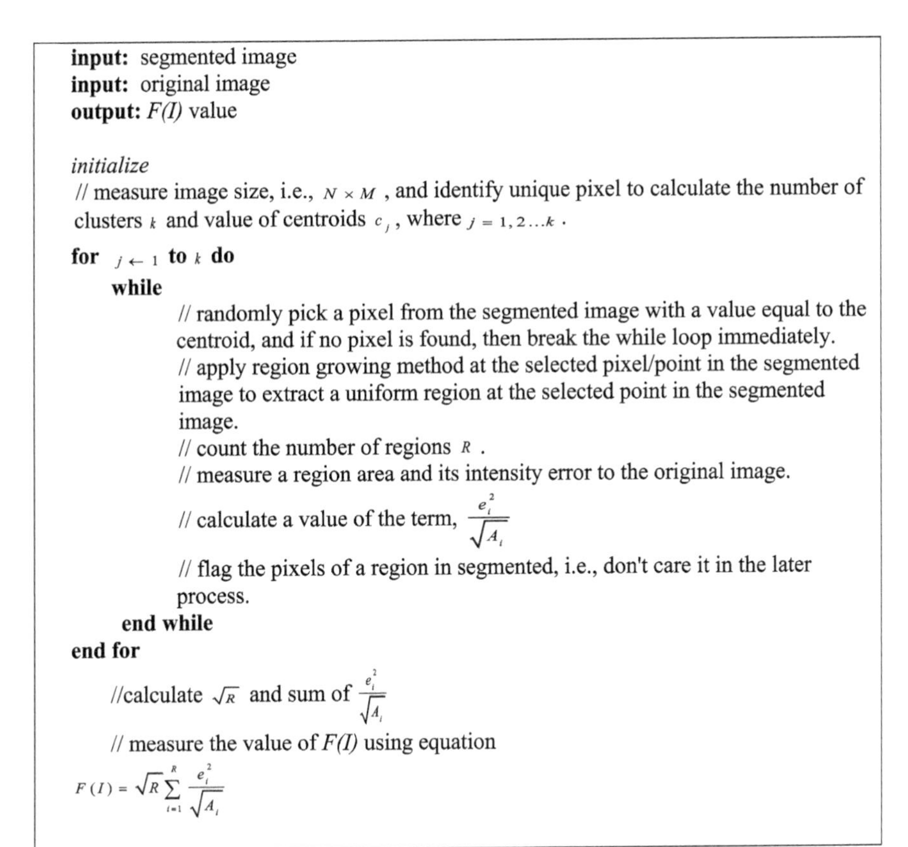

input: segmented image
input: original image
output: *F(I)* value

initialize
// measure image size, i.e., $N \times M$, and identify unique pixel to calculate the number of clusters k and value of centroids c_j, where $j = 1, 2 \ldots k$.

for $j \leftarrow 1$ **to** k **do**
while
// randomly pick a pixel from the segmented image with a value equal to the centroid, and if no pixel is found, then break the while loop immediately.
// apply region growing method at the selected pixel/point in the segmented image to extract a uniform region at the selected point in the segmented image.
// count the number of regions R.
// measure a region area and its intensity error to the original image.
// calculate a value of the term, $\frac{e_i^2}{\sqrt{A_i}}$
// flag the pixels of a region in segmented, i.e., don't care it in the later process.
end while
end for
//calculate $\sqrt{R}$ and sum of $\frac{e_i^2}{\sqrt{A_i}}$
// measure the value of *F(I)* using equation

$$F(I) = \sqrt{R} \sum_{i=1}^{R} \frac{e_i^2}{\sqrt{A_i}}$$

Table 4.4 Average *F*(*I*) value for 103 benchmark images after applying the clustering techniques

	Average *F*(*I*) value (1x10[7]) for different k			
Techniques	k=3	k=4	k=5	k=6
KM	17.020	19.953	20.146	20.677
FCM	18.243	20.465	21.825	21.958
MKM	19.658	22.999	26.898	26.993
AMKM	18.064	19.612	21.246	21.326
AFMKM	21.329	21.937	22.715	22.093
AFKM	129.60	113.83	131.60	25.890
OKM	**16.471**	**18.893**	**19.964**	**20.053**
EMKM-1	19.571	21.937	22.494	22.098
EMKM-2	17.862	21.028	21.486	21.767
ORFCM	17.480	19.885	21.668	21.642

Based on the tabulated information in Table 4.4, the best value of *F(I)* for the different clusters is produced by the OKM clustering technique. Similarly, the ORFCM produced the best results (smaller value) in the fuzzy-based clustering techniques category. The smaller value indicates the segmentation of images into homogenous regions with a high similarity index.

4.1.2.5 F′ (I)

F′ (I) is an improved version of the *F(I)* function. Unlike *F(I)* function, the *F′ (I)* penalized the segmentation with many small regions of the same size. The size of regions is the key factor that can fairly rank the homogeneity criteria of segmentation. Therefore, the term $\sqrt{\sum_{A=1}^{\max}\left[R(A)\right]^{1+1/A}}$ offers more impact of a small region on the results. If fewer small regions are segmented, then the newly introduced term's value is equal to $\sqrt{R}$ and *F′ (I)* is performed like *F(I)* function. The reformed evaluation function is defined as:

$$F'(I)=\frac{1}{1000(N\times M)}\sqrt{\sum_{A=1}^{\max}\left[R(A)\right]^{1+1/A}}\times\sum_{i=1}^{R}\frac{e_i^2}{\sqrt{A_i}} \tag{4.6}$$

The *F′ (I)* function is implemented as follow:

input: segmented image
input: original image
output: *F′ (I)* value

initialize
// measure image size, i.e., $N\times M$, and identify unique pixel to calculate the number of clusters k and value of centroids c_j, where $j=1,2..k$.
for $j\leftarrow 1$ **to** k **do**
while
// randomly pick a pixel from the segmented image with a value equal to the centroid , and if no pixel is found, then break the while loop immediately.
// apply region growing method at the selected pixel/point in the segmented image to extract a uniform region at the selected point in the segmented image.
// count the number of regions R, and measure $[R(A)]^{1+1/A}$
// measure a region area and its intensity error to the original image.
// calculate a value of the term, $\frac{e_i^2}{\sqrt{A_i}}$
// flag the pixels of a region in segmented, i.e., don't care it in the later process.
end while
end for
//calculate $\sqrt{\sum_{A=1}^{\max}[R(A)]^{1+1/A}}$ and sum of $\frac{e_i^2}{\sqrt{A_i}}$
// measure the value of *F(I)* using equation
$F'(I)=\frac{1}{1000(N\times M)}\sqrt{\sum_{A=1}^{\max}[R(A)]^{1+1/A}}\times\sum_{i=1}^{R}\frac{e_i^2}{\sqrt{A_i}}$

According to the tabulated results for clustering techniques in Table 4.5, the best value of $F'(I)$ for different cluster numbers is produced by the OKM clustering technique. Similarly, the ORFCM produced the best results (smaller value) in the fuzzy-based clustering techniques category. The smaller value indicates the segmentation of images into homogenous regions with a high similarity index.

4.1.2.6 Q(I)

The function $Q(I)$ is another modified version of $F(I)$ that evaluates the clustering technique based on the homogeneity of regions, their similarity intensity error, and the size of the regions (Borsotti et al., 1998). As compared to convention function $F(I)$, the $Q(I)$ greatly considers the size of small regions while computing the results. This evaluating function is mathematically defined as:

$$Q(I) = \frac{1}{1000(N \times M)} \sqrt{R} \sum_{i=1}^{R} \left[\frac{e_i^2}{1 + \log A_i} + \left(\frac{R(A_i)}{A_i} \right)^2 \right] \tag{4.7}$$

where $1/1000(N \times M)$ is a normalizing factor and $N \times M$ is the size of the image. Whereas the $R(A)$ is the number of regions having area A and $R(A_i)$, the number of regions having area A_i in the function.

Table 4.5 Average $F'(I)$ value for 103 benchmark images after applying the clustering techniques

	Average $F'(I)$ value for different k			
Techniques	k=3	k=4	k=5	k=6
KM	19.452	29.545	36.546	44.495
FCM	21.002	30.198	41.660	49.778
MKM	23.486	38.573	55.664	66.755
AMKM	20.801	28.180	39.465	47.010
AFMKM	23.623	31.053	38.347	45.155
AFKM	207.47	157.55	194.56	657.61
OKM	**17.696**	**27.108**	**35.546**	**42.268**
EMKM-1	23.435	33.646	43.913	49.926
EMKM-2	19.668	32.029	40.469	48.378
ORFCM	20.2433	29.683	42.386	49.432

input: segmented image
input: original image
output: *Q(I)* value

initialize
// measure image size, i.e., $N \times M$, and identify unique pixel to calculate the number of clusters k and value of centroids c_j, where $j = 1,2..k$.
for $j \leftarrow 1$ **to** k **do**
while
// randomly pick a pixel from a segmented image that has a value equal to the centroid c_j, and if no pixel is found, then break the while loop immediately.
// apply region growing method at the selected pixel/point in the segmented image to extract a uniform region at the selected point in the segmented image.
// count the number of regions R, and measure $\left(\frac{R(A_i)}{A_i}\right)^2$
// measure a region area and its intensity error to the original image.
// calculate a value of the term, $\frac{e_i^2}{1+\log A_i}$
// flag the pixels of a region in segmented, i.e., don't care it in the later process.
end while
end for
// measure the value of *Q(I)* using equation

$$Q(I) = \frac{1}{1000(N \times M)} \sqrt{R} \sum_{i=1}^{R} \left[\frac{e_i^2}{1+\log A_i} + \left(\frac{R(A_i)}{A_i}\right)^2 \right]$$

According to the result generated for the clustering techniques in Table 4.6, the AFKM shows a smaller value because of dead centroid generation during the clustering process. The best value of *Q(I)* is highlighted by bolding them. Based on *Q(I)* analysis, AFMKM and OKM produced uniform segmented regions with higher similarity indexes.

Table 4.6 Average *Q(I)* value for 103 benchmark images after applying the clustering techniques

Techniques	Average *Q(I)* value for different k			
	k=3	k=4	k=5	k=6
KM	2001.26	12211.3	49382.8	112888
FCM	2040.25	12889.4	70652.0	204829
MKM	2217.80	24071.7	141857	333357
AMKM	2191.53	10281.9	65727.3	206732
AFMKM	**1355.15**	**9288.63**	**31316.0**	116325
AFKM	3027.52	**1639.44***	**5698.71***	**43040.4***
OKM	1471.09	10533.9	37135.0	**104516**
EMKM-1	2434.62	15739.9	84585.5	190539
EMKM-2	1682.25	15081.1	55717.4	157287
ORFCM	2180.09	15203.5	93469.7	184625

*The smallest *Q(I)* value for the algorithms that fail to segment the images into a defined number of clusters

References

Borsotti, M., Campadelli, P., & Schettini, R. (1998). Quantitative evaluation of color image segmentation results. *Pattern Recognition Letters, 19*(8), 741–747.

Liu, J., & Yang, Y.-H. (1994). Multiresolution color image segmentation. *IEEE Transactions on Pattern Analysis and Machine Intelligence, 16*(7), 689–700.

Martin, D., Fowlkes, C., Tal, D., & Malik, J. (2001, July). A database of human segmented natural images and its application to evaluating segmentation algorithms and measuring ecological statistics. In *Proceedings Eighth IEEE International Conference on Computer Vision. ICCV 2001* (Vol. 2, pp. 416–423). IEEE.

Pakhira, M. K., Bandyopadhyay, S., & Maulik, U. (2004). Validity index for crisp and fuzzy clusters. *Pattern Recognition, 37*(3), 487–501.

Siddiqui, F. U. (2012). Enhanced clustering algorithms for gray-scale image segmentation. Master dissertation, Universiti Sains Malaysia.

Xie, X. L., & Beni, G. (1991). A validity measure for fuzzy clustering. *IEEE Transactions on Pattern Analysis and Machine Intelligence, 13*(8), 841–847.

Index

F.U. Siddiqui, A. Yahya, *Clustering Techniques for Image Segmentation*,
https://doi.org/10.1007/978-3-030-81230-0